Praise for *Permaculture for the Rest of Us*

Permaculture is personal journey to a  generating lifestyle. In *Permaculture fo* more skillfully weaves a thorough e; principles and practices into her own. .... land. With wit and charm she draws on her experiences homesteading in a challenging climate. She deftly illustrates how she and her family have truly created abundance while embracing the natural world around them. The result is a wealth of information, advice and inspiration.

— Darrell Frey, Author of *Bioshelter Market Garden: A Permaculture Farm*

Self-reliance is a choice today, and a difficult one, but it won't be a choice in the future. Jenni Blackmore writes with heart about her family's crawl toward living in harmony with nature. Sharing her hard-won lessons in permaculture with verve and intimate warmth, the author reveals the human side of building a new way of life: food, seeds, animals, weather and setbacks. Genuine, pithy, and filled with practical tips to encourage and guide the reader, the book also expands our knowledge of coastal ecosystems.

— Peter Bane, author, *The Permaculture Handbook: Garden Farming for Town and Country* and publisher, *Permaculture Activist* magazine

A welcome breath of fresh air, Jenni Blackmore's *Permaculture for the Rest of Us* is exactly what the title suggests: an enthusiastic, user-friendly guide to ecologically sensitive homesteading, using permaculture principles, for those who don't happen to have the money and opportunity to buy five or ten acres of perfect farmland. For those who want to grow their own food without pesticides or the other problematic features of industrial agriculture—and these days, it's hard to think of a more useful step toward personal sustainability—this is an excellent guide.

— John Michael Greer, author, *Green Wizardry: Conservation, Solar Power, Organic Gardening, and Other Hands-On Skills from the Appropriate Tech Toolkit*

As Canadians change their approach to the garden, Jenni Blackmore is leading the way. It is not that nature has changed but our attitude towards her is. Blackmore shows us how to embrace the lead that nature provides us with, in an approach that is as much storytelling as it is a guide to self-sufficiency. I am not sure if this book is more useful in the magazine rack, next to the almanac where I can pick it up and gather some nuggets of knowledge that will make me a better gardener, or next to my fat, cushioned reading chair where I enjoy a good story. We need a special category for this book.

— Mark Cullen, markcullen.com

For most, permaculture can seem like a complex philosophy too daunting to make part of our everyday reality. In *Permaculture for the Rest of Us*, Jenni Blackmore masterfully distills the concepts and principles of permaculture in such a way that the reader is convinced to put theory into practice. Both optimistic and realistic, *Permaculture for the Rest of Us* is a rare combination that provides both the how and why of creating a simpler life while fostering a deeper connection with nature, with the Earth, and with each other.

— Av Singh, Just Us! Centre for Small Farms

Permaculture
for the rest of us

Permaculture
for the rest of us

ABUNDANT LIVING
ON LESS THAN AN ACRE

Jenni Blackmore

new society
PUBLISHERS

Cover design by Diane McIntosh.
Farm painting—Jenni Blackmore. Barn/wheelbarrow: © iStock: naffarts

Printed in Canada. Second printing May 2021.

New Society Publishers acknowledges the financial support
of the Government of Canada through the Canada Book Fund (CBF)
for our publishing activities.

This book is intended to be educational and informative. It is not intended to serve as a guide.
The author and publisher disclaim all responsibility for any liability, loss or risk that may be
associated with the application of any of the contents of this book.

Inquiries regarding requests to reprint all or part of *Permaculture for the Rest of Us*
should be addressed to New Society Publishers at the address below.
To order directly from the publishers, please call toll-free (North America)
1-800-567-6772, or order online at www.newsociety.com

Any other inquiries can be directed by mail to:

New Society Publishers
P.O. Box 189, Gabriola Island, BC V0R 1X0, Canada
(250) 247-9737

LIBRARY AND ARCHIVES CANADA CATALOGUING IN PUBLICATION

Blackmore, Jenni, author
Permaculture for the rest of us : abundant living on
less than an acre / Jenni Blackmore.

Includes bibliographical references and index.
ISBN 978-0-86571-810-4 (paperback)   ISBN: 978-1-55092-607-1 (ebook)

1. Permaculture.   I. Title.

S494.5.P47B63 2015   631.5'8   C2015-905246-7

New Society Publishers' mission is to publish books that contribute in fundamental ways to
building an ecologically sustainable and just society, and to do so with the least possible impact on the environment, in a manner that models this vision. We are committed to doing
this not just through education, but through action. The interior pages of our bound books
are printed on Forest Stewardship Council®-registered acid-free paper that is 100% post-con-
sumer recycled (100% old growth forest-free), processed chlorine-free, and printed with
vegetable-based, low-VOC inks, with covers produced using FSC®-registered stock. New
Society also works to reduce its carbon footprint, and purchases carbon offsets based on
an annual audit to ensure a carbon neutral footprint. For further information, or to browse
our full list of books and purchase securely, visit our website at: www.newsociety.com

Contents

Dedicated to my wise, ever-patient husband Calum,
and my sons, Nikolas, Jason and Martin
who all did their share of digging, shovelling and hauling.
Without you, QuackaDoodle Farm could not have happened.

Introduction

Permaculture is my passion and, as with all passions, it dominates my lifestyle here on this small island just east of Halifax, Nova Scotia. With the possibility of a food crisis looming in the near future I believe it is essential to produce as much of our own food as possible. On QuackaDoodle Farm we try to do this, making the best use of limited resources by employing the principles of permaculture.

For those who may not be conversant with permaculture, it is a multi-faceted set of principles which were devised in Australia by Bill Mollison and David Holmgren. Although these principles can be applied to any system, large or small, they are especially suited to homesteading and living a fully sustainable lifestyle.

I don't think it's possible to script a single, succinct statement to adequately describe permaculture. Many have tried and indeed there are many wonderful words written about this sustainable system of living. However, an exaggerated brevity can engender vague, esoteric statements which might alienate rather than elucidate, making permaculture sound almost ethereal and not for those of us who like to call a spade a spade and to dig with one.

Alternately, a comprehensive description can seem equally overwhelming because of the apparent complexity inherent in all natural systems. In fact, permaculture is a paradoxical mix of complex simplicity. It employs naturally existing systems to produce maximum yield with minimum expenditure, more for less in other words. Minimum expenditure means using less resources which translates into less energy used, which in turn translates into a reduced carbon footprint. Permaculture principles exemplify win-win and represent, I believe, our best hope to right many of the wrongs that have been done to our beautiful planet.

It's not my intention in this book to cover every detail of every aspect of permaculture. That would make for a very long book and besides, there are several exceptionally good books already written which do just that. They tend to describe the design process, starting from A and working through to Z, more often than not using techniques appropriate for locales far more fertile and hospitable than the places many of us have the privilege of calling home. I want to share my story with those who dwell in these less benign places, who can't really be blamed for feeling just a little envious, and perhaps downright discouraged, when reading descriptions of lush garden growth in climate zones eight or nine.

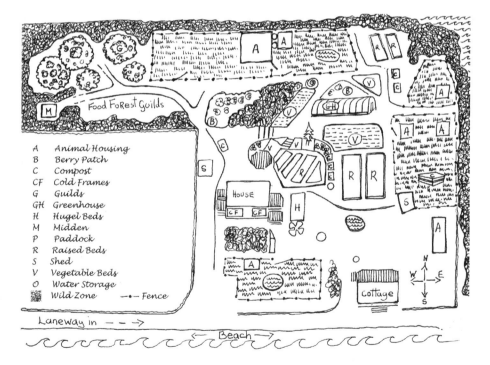

A Animal Housing
B Berry Patch
C Compost
CF Cold Frames
G Guilds
GH Greenhouse
H Hugel Beds
M Midden
P Paddock
R Raised Beds
S Shed
V Vegetable Beds
O Water Storage
Wild Zone   —·— Fence

* A basic "floor plan" of QuackaDoodle Farm.

I strive towards sustainability on a windswept island, surrounded by the North Atlantic. The soil is poor, drainage is worse and I have never once considered growing bamboo to stake my plants. The principles of permaculture were devised in Australia, where tropical plants are the norm, but I promise I will not once mention growing banana palms to thatch a shade-room! This book is about permaculture northern-style, with a few additional challenges thrown in.

My purpose here is to write an encouragement manual, an *if we can do it then for certain you can* kind of book, a book that might save others from getting bogged down by the same mistakes we made and which simplifies and elevates permaculture methodology to its rightful status. I'd also like it to be an interesting and inspiring read... and if it can elucidate some of those tricky verses from the Book of Revelation, well... just kidding! I know my limitations. The more we strive to live in harmony with the natural world the more I realize that everything requires a fine balance, including a book such as this. While not wanting it to read like a text book, I do want to supply enough concrete information to facilitate success. Much of this information will be condensed into sidebars, the "executive notes" for those who might want to skip through the anecdotal, more scenic route and parachute right into the facts. The photographs were all taken here on QuackaDoodle Farm and who knows, perhaps for some these might be inspiration enough. Whether it's a speed read during the first heady days of spring planting or leisurely dreaming on a cold winter's afternoon, read on. And enjoy!

— Jenni Blackmore
QuackaDoodle Farm
Seaforth, Nova Scotia

Slug Wars or How it All Began

Our pathway to permaculture was not planned. We were more like accidental tourists wandering into unknown territory, spurred on by nothing more auspicious than slugs. I'll get back to them in a minute but first I want to lead you back to the very beginning of the path. Fortunately some of the groundwork, let's call it the brush cutting, had been done years earlier when, as a single parent, I built the house I now share with my husband Calum. My budget was very tight and getting the project off the ground required living onsite in what, to be kind, we'll call a cottage. My sons and I lived there through the coldest Nova Scotia winter in seventy years, with no plumbing and no insulation.

Over time the roof developed a demonic system of leaks that were unstoppable. Melting snow around the uninsulated stove pipe created a deluge that required full sized garbage cans, not merely cooking pots or dish pans, to prevent an indoor ice rink from forming on the cottage floor. Technically there was no indoor plumbing despite the ample supply of running water!

Digging the well cover out of a snow drift so that it was possible to chop a bucket-sized hole in the ice lost its appeal after day one and the garbage can collection system soon became our water source of choice. It didn't take much figuring to realize that rainfall

✳ This is what winter looks like in Nova Scotia.

purposely directed into an appropriate storage tank could easily supply an average household, or so I thought. However, I still felt a little embarrassed explaining my requirements to the form builder who was preparing for the footings and basement to be poured. He didn't seem the least surprised. "So you want a cistern, Lady? Years ago all the old places had them."

A cistern! It was the first time I'd heard the word and it sounded so sweet to my ears. It validated what I feared might prove to be a crazy and unworkable scheme, one that I was hoping would help us avoid the potential hazards of depending on a shallow well that gathered only very poor quality water. The 18′ × 10′ × 5′ cistern is built into one corner of the basement and holds six thousand, six hundred and fifty gallons of good clean water. With typical rainfall it tends to remain at least half full most of the time. Actually seeing and being able to gauge the availability and usage of water caused a subtle shift in perception and gave me a clearer insight into the effective management of resources. Step One along the permaculture path! I just didn't realize it at the time.

The ground here is very poor, pretty much solid clay topped with an inch or two of highly acidic top soil. It's enough to support stunted spruce growth, which isn't saying much as it is not unusual to see a ragged spruce anchored to the side of a rock face. In its natural state the soil certainly wasn't up to nourishing my illusive dreams of growing vegetables. After a couple of dismally disappointing growing seasons I embarked on some half-hearted and poorly informed attempts to build soil. Poorly informed? Well, for instance, I didn't understand that the stinking piles some gardeners raved about were not giant fly and vermin lures but in fact represented one of the essentials my soil lacked: *compost!* It was a word I barely knew the meaning of, never mind its importance.

Fortunately I am within a hundred feet of a constant supply of seaweed—one of the advantages of living along the seashore. The bags of leaves left curbside in the city also provided "easy pickin's" and over time these organic additions began to work their wonders in my vegetable plot. I wasn't fully conversant with the term "organic material," I just knew these freebies were in line with my very limited budget and the need to bulk-up my soil. Did I consider myself to be an organic gardener back then? Absolutely not. I was just a low-income mom being ruled by necessity.

Similarly, budgetary restraints—or encouragements, depending on your perspective—dictated the size of house I built. By current standards it might be described as modest, although I prefer to think of it as more-than-adequate. The living room, kitchen and dining nook are open plan, allowing for excellent air flow from the centrally located wood stove. The heat also rises up the open stairwell to the upstairs bedrooms, keeping them at perfect sleeping temperature. If I hadn't spent that bitterly cold winter in an uninsulated, one room cottage prior to building, I doubt that I would have designed this house as effectively, considering efficient heat distribution to be of primary importance.

The fact that the house is oriented towards the south and takes full advantage of passive solar heat was pure luck, dictated by the uninterrupted ocean view. "Solar" was another unfamiliar word back then and while it certainly didn't feel like I was starting to live a "Permie's dream" I realize now that I was, in fact, assembling the foundation blocks for just such an experience.

* Passive solar heating in the house is a fortunate happenstance of this view to the South.

Around the time my veggie plot was grudgingly pushing up the occasional taster of spinach leaves, and pea shoots that actually produced pods, I met my dear husband-to-be, Calum. As a biologist he was light years ahead of me when it came to gardening. He even produced compost that was light and fluffy and more importantly, didn't smell! With our combined efforts the garden finally began to flourish...and so did the slugs. They were a plague of biblical proportions. Halved grapefruit skins, copper mesh, crushed clam shells and beer enough to drown in; we tried them all, and all with a certain measure of success, but never enough. Then we heard about the sure fire silver bullet, chickens. Some chickens maybe, but the heritage breeds we opted for gave us the malevolent eye and the "Cluck-cluck! You expect us to eat what?"

By the time we realized that our slug elimination plan number six (or was that seven?) was definitely not working we had already consumed our first fresh laid egg. There was no going back. The chickens had arrived to stay. Thanks to them and their prodigious output of manure, the veggie plots are now edible jungles by early August and they supply us with most of the fresh produce we need from late May through to December. This in itself is quite an impressive feat, because we really do like our vegetables.

✽ "Eat what? We don't do slugs!"

There is another less concrete, but equally important benefit to keeping chickens, which is more difficult to define as it speaks to the soul, to our intrinsic need to be integrated more fully with the natural world. How do chickens do this? I don't quite understand, but on a blustery winter's morning, when the last thing I feel like doing is trudging out to the chicken coop, their passive cooing as they peck at their breakfast nurtures something primal within me and reaching under warm feathers to retrieve a fresh laid breakfast gift satisfies places other than the stomach.

Calum's way of approaching a new project is to read up on all the available information first. This methodology provides an excellent counterpoint to my more impetuous way of plunging in with both untutored feet and hopefully learning from my

A successful permaculture plan encourages multi-functionality. Elements are chosen and positioned to perform more than one primary purpose in a self-sustaining way, while at the same time integrating fully with other elements. Chickens exemplify this principle perfectly. Because of their multi-tasking potential, they are iconic to permaculture. Their primary purpose is to supply eggs and perhaps meat, but in a garden situation they also aerate and fertilize the soil while consuming harmful bugs and plucking out the odd weed. Given the opportunity (i.e., with a rooster present) they complete their self-sustaining cycle by hatching their eggs to produce the subsequent generation of egg layers and garden helpers. To elaborate further, fresh free-range eggs and young chicks, especially heritage breeds, are an easy sell and this "egg money" provides an external cash flow. A simple *hen-poster* (a box containing organic waste that allows easy access for chickens) and a regular supply of kitchen scraps is all chickens need to very quickly produce great compost with their constant scratching and manuring. The best thing is that they do all this for, yes, you guessed it...chicken feed.

mistakes—eventually. Through this combined approach, and the acquisition of several of the excellent books available, we have become quite savvy in several areas of sustainable living. It often seems that one step along the permaculture way leads quite logically to a couple of more steps, but not necessarily in a straight line. As with keyhole perimeters that help expand the accessible area of a garden, so too this meandering learning curve further expands the breadth of knowledge that it is possible to gain.

Keyhole gardens have become an icon of permaculture because they provide a clearly visual example of efficient use of space. They also maximize on the positive dynamic of edges. Edges, areas where two distinct entities meet, are dynamic because the attributes of each entity are able to intermingle. For example, geographically, a border town will be influenced by the cultural aspects of its neighbour such that, while still maintaining its own cultural identity it will form a vibrant mix of cuisine, music, architecture, etc. Similarly, in the garden, edges draw from both sides. A path alongside a garden bed allows for additional sunlight to reach plants. It might also direct run-off into the garden and certainly it provides easy access for care and harvesting. A scalloped, meandering line is of course much longer than a straight line and so provides more edge dynamic, while the quintessential central pathway gives maximum access with a minimum loss of growing area.

The elimination of slugs, as another for instance, led us to the acquisition of chickens which in turn developed our interest in heritage breeds, which lured us into a breeding program for critically endangered Pilgrim geese which have, in true permaculture fashion, eliminated our need for lawnmower and grass trimmer. At the time of this writing we have a profusion of goose eggs in incubators, as well as in the nests outside, which are being proudly protected by parents that, this time last year, were pecking their way out of carefully tended eggs. At this rate they won't be on the endangered list for long!

We also raise Khaki Campbell and Runner ducks, both noted for their phenomenal egg laying capacity. Better yet, they love to eat slugs! As well, we have started raising Beltsville White turkeys, which are worth keeping for their entertainment value alone. As an added bonus they dress down to make the best Christmas dinners ever, without a doubt.

And while on the subject of tasty dinners I mustn't forget the Californian meat rabbits who share the duck barn along with the geese. After harvesting our first litter recently, we realized that we are moving ever closer to our target of becoming at least 60 percent self-supporting.

I guess you could say that we are gradually retrofitting our lives to align with permaculture ideals. No doubt it would be easier to start from square one, with a perfectly positioned piece of virgin territory, but I doubt that many ever have that luxury. And

✳ Ignore the cuteness and never, ever call them "Bunnies."

besides, every natural environment surely has its own particular challenges which will dictate the "zones and guilds" an aspiring permaculturist needs to establish.

When I first developed an interest in permaculture and started amassing a library of the great books available I'd gobble up the first few chapters but begin to lose my appetite when it came to zones and guilds, in part because the diagrams looked a little overwhelming in their apparent complexity. I was tempted to—okay, I did—hurriedly scan through those chapters so I could get back to the growing and harvesting parts that interested me the most. My abject fear of all things mathematical was rekindled by any mention of degrees of slope, angle of sunlight, or exactitude of compass points. In actual fact, the concepts of zones and sectors are based in common sense and are essential to any successful permaculture plan.

Zones range from one to five, from the most often to the least often visited. For example, I don't want to have to charge to the far end of the garden for some fresh basil, with my sauce already bubbling and company due to arrive in five minutes. Therefore zone one includes herb pots on the deck by the kitchen door. Potatoes, once planted, require almost no attention until harvest time, so they will be best placed in zone three.

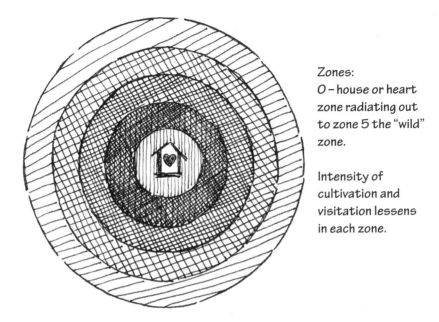

Zones:
0 – house or heart zone radiating out to zone 5 the "wild" zone.

Intensity of cultivation and visitation lessens in each zone.

✳ This is a theoretic diagram. In actuality zones spread and flow more like broken egg yolks in a fry pan as they are influenced by their topographic environment.

Sectors are like slices of a pie-chart that clearly define the sunniest spots, the wind tunnels, any natural water courses and so on. Once these characteristics have been itemized it's much simpler to take full advantages of their attributes and take steps to minimize any negative effects. Having them drawn out on paper helps solidify the existence of these invisible boundaries.

Every homestead, regardless of size, will be unique when laid out to incorporate permaculture ideals. This is one of the many wonderfully satisfying aspects of living in harmony with the natural world although, as with any retrofit, things don't happen overnight. Patience is a virtue that doesn't get mentioned often enough as a necessary requirement in many of the informative books already written on attaining the dream-state of sustainable living. On the other hand, even though things might not always zoom along on the permaculture way, they often seem to have a knack of integrating and developing very smoothly.

A few years ago hurricane Juan ripped through Nova Scotia. If I was a paranoid type I'd believe it specifically targeted our property, leaving in its wake a gnarly, impassable wreck of fractured and uprooted trees. For a while we tried the *"I can't see you"* and the *"If we ignore it, it will go away"* approaches, which were as successful as might be expected. Finally we began the painstaking (read *aching muscles*) job of first removing the debris and then attempting to reclaim the land.

After weeks, then months, of exhausting and dispiriting work it seemed that progress was virtually a non-happening. Because progress was so gradual any proof that it actually was happening tended to jump out unexpectedly; sudden realizations at the sight of brilliant orange pumpkins flourishing on what had been ravaged forest floor or awakening moments ignited by the strike of sunlight on mammoth sunflowers towering above broken tree stumps. It was at times like this that we forgot all the discouragement and

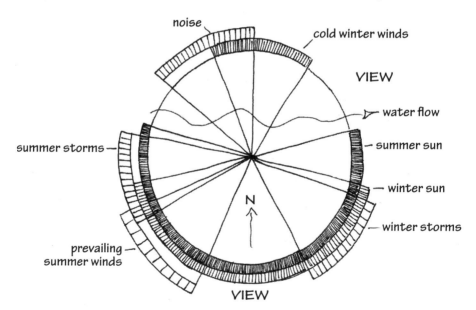

★ This particular sectors diagram is a very basic analysis of some of the prevailing elements which influence QuackaDoodle farm. It does not take into account the microclimates and other deviations which are created by trees and buildings.

✳ Cleaning up after the hurricane was a daunting task.

instead allowed ourselves to be amazed at how well everything was coming together. As time progressed these moments became more frequent and now, looking around our fertile little haven, it's sometimes hard to remember how broken and barren it was back then.

The sunflowers mentioned above were planted in used feed sacks interspersed around the berry patch. The ground was so root riddled and virtually impenetrable in places that it was much easier to "plant up" and mulch heavily rather than try to dig down. The roots left in the ground gradually rotted and now help to maintain moisture while also creating an explosion of beneficial microbes and bacterial action. Over the past couple of years this piece of ravaged ground has become amazingly fertile and supplies liters of lush berries. It also exemplifies one of the mainstays of permaculture philosophy: there are no problems in permaculture, only creative solutions.

The location of pathways surrounding the various plots was dictated to quite a degree by the location of larger immovable stumps that will obviously take more than a couple of years to rot away. As it happens, this seemingly random layout has created a pleasing yet efficient organization of space. The position of the greenhouse, for example, was determined by the location of a rel-

✳ Squash, beans and sunflowers thriving on reclaimed forest floor.

atively unobstructed space. It now effectively blocks the prevailing ocean breeze while still allowing the sunlight to pass through and has provided a snug microclimate for the berry patch to the north of it.

As well as blackcurrants and blueberries I planted a couple of Northern Kiwis and a couple of grape vines. My plan is to create an arbor type of setting that will disguise the netting used to protect the blackcurrants and blue berries from the birds. They don't share very well when it comes to berries!

There are several Indian Pear trees on the property. These never did much when competing with the spruce trees but are now re-

✳ The perimeter of the berry patch is dictated by immovable stumps.

sponding well to our nurturing and hopefully will help satisfy the birds' needs. Maintaining enough wild area to support the needs of wildlife is one of the rules of permaculture and we have left several pockets of wild growth which have regenerated with an amazing diversity of native species.

The ducks and geese also benefited from the devastation of Hurricane Juan (it's an ill wind etc.) as they now have a large open corral with their own 10' by 15' barn, which they share with the rabbits. Rainfall runs off the metal roof, down the eaves-trough and is collected in 45 gallon barrels which gravity feed the duck pond. Very simple, very effective. This brings up another permaculture aim, which is to generate maximum gain from minimum expenditure. As an example, the non-permaculture way to fill the duck pond might be to use an electric pump and well water. The effect would be the same but the water used might have been better conserved for other use and the electric pump would obviously add to our carbon footprint.

As already mentioned, the layout of our design was definitely influenced by the immovable stumps and the need to construct pathways around them. As it happens, I don't believe we could have done it half so well sitting at a desk with a blank sheet of

* This Indian Pear tree graces the wild zone every spring.

paper and no restraints. This exemplifies another law of permaculture which outlines the necessity of an intimate knowledge of the land you intend to unite with. It is essential to be aware of every bump and undulation, every wild flower, every shrub. Walking and working on the land, even in the early stages of clean-up, certainly created in us a much better understanding of what we had to work with.

One of the few plants that really seemed to thrive in those early days was comfrey. It is a terrific source of nitrogen, whether as compost or simply cut and laid between rows as a green "manure" mulch, but at the time I saw it only as a weed that needed to be removed. More about comfrey and other naturally occurring plants later, but for now I must just mention how glad I am that all my early attempts to eradicate it failed. If I had been successful I would have lost many benefits and certainly would have been in conflict with another permaculture practice—encouraging the preservation and diversity of nature.

Similarly, yellow irises thrive here to a height of nine or ten feet and in the spring several kinds of wild ferns grace the property with their elegance. I have purposely encouraged clumps of them in various spots where they originally chose to grow. Ferns don't much like being moved but will accept being re-located as long as it is done as early in spring as possible, with the least amount of root damage. Hosta lilies, although not naturally occurring, were one of the few perennials willing to stick around during my early gardening days and they make great natural retaining walls on sloping banks. Learning to work with what we have has made life so much eas-ier. *Encourage what wants to stay—let the rest go away.* That has been my motto for a while now.

I don't believe that any knowledge-able person in their right mind would have chosen this location to start up a micro-farm or even a serious garden.

✳ This slender young comfrey shoot might look innocent enough but comfrey grows in dense, tenacious clumps that can be several feet high. Even so, no Permie Plan should be without these plants, because of their superior performance as soil enhancers.

The North and South boundaries of the property merge into shore line and such close proximity to the ocean means that tempera-tures in the growing season are always cooler than on the main-land and more often than not a wind of some description will be blowing. The existing soil can only be described as being poor to terrible and the solid clay substrata creates its own set of drainage problems. And, it's not like we have "acreage." We have just about one acre, that we are successfully turning into a micro-farm which, if all goes according to plan, will eventually make us around sixty per cent self-sufficient.

Other than the view and the tranquillity it has to offer, this place could not be described as the optimum choice for a self-sufficient lifestyle, but it has led us and taught us how to discover what is possible and that of course is half the thrill. And yes, life is a thrill, in strange unexpected ways. Just as the land itself dictates how it will be used, a permaculture plan will direct, at least to a certain extent, how the lives of those who practise it will be lived. We try not impose our will but rather to integrate our living as just another dimension of the great and glorious cycle we all participate in.

If this pathway into permaculture seems somewhat circuitous and meandering so far, well, that's a good thing, because in fact the natural world is not structured in straight lines. Each healthy system rolls back on itself, forming its own sustainable circle while also connecting to many other circular systems, which in turn connect.... This intricate interconnectedness makes it difficult, near impossible, to stay on a linear path while also stating the actual facts. And besides, I don't think the amazing beauty of natural design should be sliced and diced too finely. Attempts to isolate specific elements to the detriment of others are greatly responsible for the devastating imbalance that is presently endemic to most of the world's agriculture. Having said all that, I do intend to keep subsequent chapters more focussed on the particular topics under discussion. I promise.

The Nitty Gritty
on Building Dirt

It's easy enough to draw well thought out plans, to map how the zones and guilds will merge and to visualize how satisfying it will feel to gaze out over beds of healthy vegetables—but without good soil these dreams will not come true. Good soil is absolutely essential. Fortunately, with some scavenging, a bit of patience and a certain amount of shovelling, it can be "constructed." The naturally occurring "soil" on our little farmstead is virtually solid clay and I would say about 80% of the content of the soil that my lovely vegetables grow in has been imported from elsewhere, in the form of compost, manures and various other organic materials.

Descriptions of soils usually refer to the clay or sand content. At its worst clay does not drain, is too dense for fragile roots hairs to penetrate and is often alkaline. Sandy soil drains too quickly, allows any existing nutrients to be leached away and leaves plants dried out and undernourished. These are the extremes. I work with the first type. Most soils tend towards one or the other. Moistened clay soil will tend to ball up when rolled between thumb and forefinger whereas sandy soil is more likely to crumble. Organic

Soil Types

There are three main components to soil; sand, silt and clay. The ideal ratio is 40 per cent sand, 40 per cent silt and 20 per cent clay. Of course this perfect mix is seldom present. Soil that is too sandy does not hold on to nutrients or moisture. Soil that has a high clay content does not drain well and often becomes waterlogged causing roots to rot. Silt heavy soil compacts easily creating poor aeration. A simple test can determine the true nature of soil.

Soil samples are best taken from about six inches below surface. A typical glass jar is filled one third with soil sample then topped up with water and shaken thoroughly. (If available a spoonful of powdered dish or laundry detergent can be added to aid dispersal of particles.) Sand will settle after one minute. Silt will take four to six hours. Clay will take up to two days. For soils high in clay or sand, the addition of organic material will help—to increase the moisture and nutrient holding capacity in sand and to add fertility and increase drainage and fertility capacity in clay.

Clay Soil
organic material
muddy water
thick silt layer
some sand
some gravel

Loam
some organic material
murky water
thin silt layer
sand particles
gravel

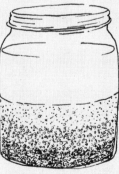

Sand
no organic material
clear water
no silt
layers of sand and gravel

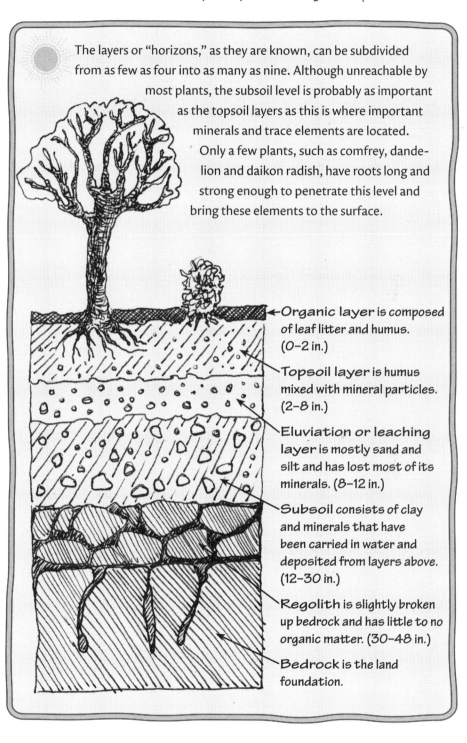

The layers or "horizons," as they are known, can be subdivided from as few as four into as many as nine. Although unreachable by most plants, the subsoil level is probably as important as the topsoil layers as this is where important minerals and trace elements are located.

Only a few plants, such as comfrey, dandelion and daikon radish, have roots long and strong enough to penetrate this level and bring these elements to the surface.

←Organic layer is composed of leaf litter and humus. (0–2 in.)

Topsoil layer is humus mixed with mineral particles. (2–8 in.)

Eluviation or leaching layer is mostly sand and silt and has lost most of its minerals. (8–12 in.)

Subsoil consists of clay and minerals that have been carried in water and deposited from layers above. (12–30 in.)

Regolith is slightly broken up bedrock and has little to no organic matter. (30–48 in.)

Bedrock is the land foundation.

material is the key to rectifying either of these problems. As mentioned previously I initially added large quantities of seaweed and leaves because they were accessible and affordable.

Seaweed is an excellent additive. My apologies to those readers who don't have the advantage of a local source but I do need to rave on about it, just a little. The way I see it, there are two broad categories of seaweed: eel grass, which washes ashore for harvesting each fall and kelp mix, which is more likely to wash onto the beaches after a storm at any time of the year. Eel grass looks like brownish-black paper that has been shredded, whereas the kelp mix is more variegated in oranges, browns and greens.

Eel grass is interesting as it has several traditional uses. Its ability to pack densely and still maintain lots of air pockets makes it an excellent insulator. It can still be seen banked around the foundations of older homes and the difference this makes on a cold winter night is really phenomenal. It can also be banked around wells and waterlines to prevent freezing and was actually sewn between sheets of brown paper and used as in-wall insulation at one time. It is also great as a medium for growing potatoes. Of course, they come out perfectly clean when harvested.

Eel grass also makes great mulch and it seems to me that slugs don't like it much, perhaps because it has a hint of salt about it and fine particles of sand embedded in its strands. There's not much that can be said against eel grass, other than it is very slow to rot down (decompose) and therefore is not a brilliant nutrient enhancer. It does, however, improve the friability of the soil over time.

The kelp mix is much quicker to decompose and is highly beneficial to the nutrient content of any soil. This mix will no doubt include rock weed, wrack, dulse, Irish moss and other sea vegetables, as well as broad rubbery ribbons of kelp. If possible, it is best gathered before it's rained on too much as some of the nutrients will leach out in fresh water, leaving the kelp whitish in colour rather than its usual rich brown or olive green. Some rain, however, helps to get rid of excess salt.

* This bed is composed almost entirely of seaweed, leaves and manure. Over time it has morphed into a highly fertile and friable soil mix.

The mix will probably have been rolled around by an angry sea and will no doubt be well and truly tangled. A shovel just doesn't work. A spading fork is better suited but a manure fork with slightly curved tines works best for lifting the mix into feed sacks or plastic panniers. Chances are it will be just hopping with tiny sand shrimp but they have a way of disappearing. As far as I know they don't harm anything in the garden and will provide further nutrient addition to the soil.

Sometimes I wonder about accumulating too much salt residue in the soil. Allowing the seaweed to be piled and rained on before adding it to the garden would prevent this, but would also result in substantial nutrient loss, so it's a trade-off. I've used a lot of seaweed straight from the beach and have not yet had a problem with excessive salt.

The alternative to slithering around on banks of kelp is to buy high potency, 100 percent organic, liquid fertilizers made from kelp concentrate, which can be sprayed directly onto leaves of

Soil science, much like quicksand, can suck in and rapidly overwhelm the unsuspecting explorer. I'm just dipping one little toe into the periphery here. It's fairly common knowledge that plants need Nitrogen (N), Phosphorus (P) and Potassium (K). On typical fertilizer labels they are listed as three numbers, such as: 10-10-10 or 20-20-20 or 20-5-5, the latter having four times more nitrogen that Phosphorus or Potassium. Trace elements are seldom mentioned, even though they are equally important to both the plant and the consumer of the plant. Trace elements often occur in minute amounts and in subtle but crucial balance. They are not readily available in the topsoil layer and need to be accessed from the subsoil clay. This is where dynamic accumulators such as Comfrey come in. They have the ability to suck up trace elements from the subsoil, making them available to more shallow rooted plants through compost, teas and mulching.

Trace elements or micronutrients, as they're sometimes referred to, can be added as rock phosphate, greensand, granite dust, etc. but since excess can be as detrimental as insufficiency I prefer to rely on dynamic accumulators and gifts from the ocean—which can be collected, or bought as kelp meal and fish fertilizer.

For example:

Comfrey accumulates N, K, Ca, Mg, F, and Si.

Dandelion accumulates P, K, Ca, Mg, Fe, Cu, and Si.

They really do deserve a lot more respect than they get!

And for anyone such as myself, who never learned or can't remember the periodic table, the chart below will be useful.

Major and Minor Elements of Interest to Gardeners

B	Boron	Mn	Manganese
Ca	Calcium	Mo	Molybdenum
C	Carbon	P	Phosphorus
Co	Cobalt	K	Potassium
Cu	Copper	Si	Silicon
Fe	Iron	S	Sulphur
Ma	Magnesium	Zn	Zinc

plants as well as around the roots. (Easier, perhaps, but not half so much fun!) Dried kelp meal is also available which can be dug in or used as a top dressing. Most natural sources of micronutrients in perfect balance have been disrupted by the use of pesticides and chemical fertilizers. Fortunately the ocean can still deliver a full range of trace elements in perfect balance. Liquid fish fertilizer is marketed in much the same way as kelp fertilizer. Both of these concentrates are commendable: kelp for soil amendment, fish for mid-season feeding.

Leaves are wonderful, and free, soil builders. They are easier to come by in the fall but in the early days of spring bags will also appear at curbside. Nutrients vary slightly from tree species to species. Oak is my favorite while poplar is my least favourite, as it always seems to come with lots of tough little branches mixed in. Some of the goodness in spring gatherings will have leached out over the winter but, safe to say, any leaves are better than no leaves. Ideally they should be shredded by running a power mower over them a few times. I don't do this. For one thing it's time consuming and honestly, boring, but my valid excuse is that it requires a noisy, smelly gas mower. Worms, beetles and other crawlies do a much better job. They take a bit longer, which is another good reason to add leaves to the ground in the fall.

As with seaweed, leaves can be used as mulch or to bulk up the organic content of soil. They can also be composted, which brings us to the third really important ingredient of good soil, compost. Just about anything organic can be composted but a typical garden compost pile will contain a wide variety of green and brown organic material including kitchen and garden waste, grass clippings, leaves, spoiled hay, manure, and if possible, a small amount of mature compost and a couple of handfuls of humus from the forest floor to ensure the presence of plenty of worms and microorganisms. (I like to think of these last two additions as "starters" that will act in the same way a sour dough starter or yoghurt culture does.) Each layer needs to be watered when added. A compost pile can be as simple as a heap of organic refuse, preferably tucked away in a distant corner of the lot, or as "high tech" as an aerated barrel that is rotated regularly and kept conveniently close to the house. Who would have thought there could be so many variations on the theme of rotting vegetation!

I had a rotating barrel once. It was not factory made and I was pleased to support local initiative at first, until I discovered the many design flaws this particular initiative incorporated. As the barrel filled up it became increasingly difficult to turn, requiring that I lean all my weight against the handle. This positioned me directly in the splash zone of malodorous fluids that slopped out with every quarter turn. Often, usually when the barrel was almost full, the hatch had a habit of bursting open during rotation, scattering slimy, half rotted contents over the rotateur (that would be me). It wasn't long before that composting system stalked down the lane on its spindly recycled iron legs to await pick-up, no doubt by some unsuspecting, compost-hungry green thumb. I'm a rabid garbage hound myself but I have to acknowledge that occasionally there is a good reason why something is put out for garbage collection or "free-cycling."

✳ This highly efficient composter is simple to construct.

＊ Each "slice" is removable and interchangeable.

＊ Clear plastic lid aids the heating up process and prevents soaking and leaching during rainy periods.

But back to compost: It does work better when contained, rather than just piled freeform. The container can be as simple as a length of snow fence or chain-link fencing, curved to form a cylinder or staked in four corners to form an open ended coral; or several recycled wooden pallets secured to form a large wooden box with plenty of aeration. I believe we have a near perfect and pleasingly simple system. It consists of three segmented boxes. The segments or slices can be lifted from one box and placed onto the next, so that as one box fills additional slices can be added to accommodate the growing pile of compost. When it comes time to turn the compost, i.e., to shovel it into the next box, the slices are removed one by one, reducing the height of the emptying box. As the well-on-its-way-to-being compost is being shovelled back and forth between the first two boxes the third box is used for fresh compostables starting the next load. The "turning" or shovelling back and forth is essential to fully aerate the pile. This encourages it to "heat up," accelerating the progression towards a rich, fluffy finished product rather than a slimy mess. Additions of fresh greenery or grass clippings will also accelerate the "heating up" process, which is important in order to destroy any unwanted seeds in the mix.

A single leaf of yarrow is said to drastically speed up the decomposition process. It grows wild and is not unattractive. This is just another ex-

✱ Yarrow has been prized since Roman times as a powerful healer. It is thought to protect surrounding plants from disease, perhaps by elevating their own resistance levels. Its straight woody stems were used by Confucius (and many others) for divination by the yarrow stalk method for consulting the I Ching.

ample of how much is lost when all naturally occurring vegetation is delegated to the *Yikes! A weed—must eradicate* list.

Of course there are many weeds that we can live without, especially in the garden, and equally certain is the fact that most weeds are incredibly vigorous and determined. In theory if a compost pile heats up as it should all seeds are destroyed but when it comes to weeds I don't push my luck. I prefer to put all weeds in the less sophisticated compost pile. This is a series of open ended corals, used for coarse materials that take longer to rot down and also used for larger quantities, such as barn clearings, that will likely be a potent mix of manure and organic bedding, guaranteed to heat up sufficiently for seed destruction.

There is indeed quite a science to composting. Whole books are written on that topic and perhaps that's one of the reasons I shied away from it for years. It just seemed way too complex. Now I wonder why I was such a stubborn convert. It isn't complicated and it isn't arduous and it doesn't need to be smelly or fly infested either…well, not very. Turning a compost pile might take ten minutes; the more often this is done the more quickly the ingredients break down. The first turn is the toughest but each subsequent turn gets easier. A fork works best at first and when it starts to seem like a spade or shovel would probably work better it's time to think about spreading this amazing nutrient source directly onto the soil.

To call compost an "amazing nutrient" is no exaggeration. There's a certain alchemy that happens when all those shrivelled peelings and sad looking leftovers start to break down. The "waste" organic material becomes alive with microbes and beneficial bacteria that create complex systems. These systems enable the growing plants to much more readily absorb the nutrients made available to them, while at the same time deterring the proliferation of harmful microbes and bacteria. This is a highly simplified version of the basic science of soil. (For more detail see Suggested Reading.)

Worms! I saved the best for last. Compost piles are the mega-cities of the worm world. They redefine the phrase "teeming with" and, of course, worms are the super heroes of soil. They aerate the soil with their complex web of tunnels, which also help give soil a sponge–like quality, enabling it to absorb water and preventing run off. Worms help to break up and mix together the various elements in the soil which they then carry from the surface down to the hungry roots in the form of worm castings (a polite way of saying worm poop). They are quite like road crews equipped with microscopic picks and shovels, except they work much harder and do way less damage. :-)

I don't think there can be such a thing as too many worms in a garden. One way to generate sizeable amounts of the amazingly rich worm castings, while at the same time instigating a population explosion, is to set up a worm culture in a typical plastic tote box. Astonishingly simple and efficient, this is a sort of compost-for-dummies approach which requires only a few start-up worms. They will reproduce and poop like crazy as long as they are given a regular diet of kitchen waste to work on. The worms to use are Red Wigglers (they are smallish, deep pink to red and yes, they wriggle) rather than the larger, paler coloured species of earthworm. Anyone who has an existing compost bin will be able to give you a handful—okay, a container full—because compost bins seem to be the equivalent of Disneyland to wigglers. It's easy enough to harvest a start-up colony as they thrive in great clumps

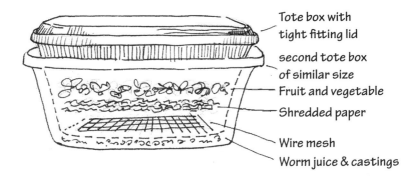

Tote box with tight fitting lid

second tote box of similar size

Fruit and vegetable

Shredded paper

Wire mesh

Worm juice & castings

✳ A typical collection system made out of two rubber tote boxes.

If you're one of those people who couldn't bait your own fish hook when you were a kid, you might want to skate over this section—but don't discount the information because worm castings and "worm juice" are like super mega nutrition for plants. It's all about red wigglers. They do have several other names but to my mind I think that one best describes them. You can have a "worm farm" under your kitchen sink (recommended, but not by me) or just about anywhere else, inside or out. In-house, a tight fitting lid is essential, otherwise imagine *Snakes on a Plane* in miniature! For more than you'll probably ever need to know about vermiculture, or "worm farming," visit: oacc.info /docs/vermiculture_farmersmanual_gm.pdf. This site really is the be all and end all source of worm farming information.

just below the top layers of compost. They are not into burrowing like a typical earthworm.

I quite enjoy turning my compost pile on a cool spring day when I want to be outside and digging even though it's too early to disturb the soil, but for true joy while turning compost chickens take the prize. In the chicken run Calum constructed what we call the Hen-poster, a square about 5' × 5' and about 18" high in which we pile green waste and the occasional bag of leaves. The chickens do a fantastic job of rooting, mixing and manuring whatever they're given, which in turn composts quite brilliantly.

Interestingly, the chemical profile of soil can be manipulated to quite a degree by what is added to it and it really is worthwhile having a soil test done by your local Department of Agriculture extension office. While commercial companies will test for heavy metal contamination and other specific problems, the Department of Agriculture will examine the quality of the soil in terms of organic content, pH balance, nitrogen, phosphorous, calcium content and so on, and they will suggest what is needed to fix deficiencies and neutralize excesses.

✳ Chickens at work transforming garden and kitchen waste into "magic" soil.

I suspected years ago that I had "sour" or acidic soil. To counteract that and because "Excess" is my middle name, I spread limestone quite heavily every year, thinking that you can't have enough of a good thing. Wrong! Balance is everything. A soil that is too alkaline encourages various blights and scourges, and some plants such as blueberries and potatoes prefer soil that is slightly acidic. Test strips are available at garden centers which will give

> Important note here that can stand re-stating: Chicken manure is wonderful for the garden but it absolutely has to be composted for a year at least. If applied "fresh" it will "burn" the plants as it is too "hot." This rule applies to all manure to a lesser degree, except for rabbit droppings which can be applied directly to the garden and in fact should not be composted as this reduces their effectiveness.

a reading on the pH balance, but not the nitrogen/phosphorous balance which is also very important.

In my opinion sawdust and woodchips should not be used on a garden, not even as mulch (except on blueberry bushes). They will add organic texture to soil but they rob it of its nitrogen because the microbes that break down the wood fibers also gobble up the nitrogen. However, a small amount of bedding from a chicken coop, which contains wood chips, can be used to improve texture because chicken poop is extremely high in nitrogen. It's important to note that fresh chicken manure applied directly to the garden will "burn" everything because it is too "hot." I should clarify here

Wood shavings need to be especially well composted as they have a very unbalanced carbon to nitrogen content (500:1 C:N). The bacteria which break down cellulose, making it available for plants to feed on, require a carbon nitrogen ratio of 30:1 so of course they're not happy in wood shavings. After they have used up the nitrogen available in the mulch they "rob" nitrogen from the surrounding soil leaving none for other plants, which therefore wither and die. Even the "stolen" nitrogen will most likely not be sufficient to meet the needs of the microbes and many of these will also perish. In addition, wood chips, because of their large surface area and due to the presence of lignin in the woody cellulose structure, are slow to decompose. A similar process can be attributed to sawdust except that it is not so pronounced because the woody fibers have already been broken down somewhat and the smaller particles create a much larger surface area which further facilitates decomposition. I know that blueberry bushes welcome sawdust as a top dressing, but even so I only tolerate saw dust and especially wood shavings near my garden if I know they have been liberally scattered with manure (preferably chicken) and then only after this mix has been well composted. I believe they are best avoided altogether.

that compost "heat" (already mentioned as beneficial) refers to actual temperature elevation, whereas "hot" manure (which might well be steamy and elevated in temperature) is extremely high in nitrogen and is also highly acidic. This is what "burns" the plants, causing them to wither and die. Once it has been aged for at least a year there's nothing like it but still, use with care.

I tend to think of nitrogen much the way I used to think of lime. More is good so even more must be better. This is not necessarily true, especially as plants such as beans can get all the nitrogen they need on their own. At the same time, there is no doubt that some of the heavy feeders, such as squash, would agree with me. This is where I introduce, with drum roll, the comfrey plant. I used to despise comfrey. It's more stubborn than I am. It is unruly, undisciplined (yes, I guess we have quite a bit in common) and it seems to thrive on being cut down, beaten back and in any way opposed. So why did I wage such a (losing) battle against it? Because I saw it as a noxious weed that was setting up to be the nemesis of my inner gardener. That was in my pre-permaculture days, before I learned that comfrey is a brilliant source of nitrogen. It also attracts bees like crazy. As an added advantage its roots can reach several feet deep into the ground, breaking up previously impenetrable clods of soil while drawing precious trace elements to the surface. These essential micronutrients can then be made available to plants with less adventurous root systems. Harvested comfrey can be added to the compost pile or used as a green mulch around plants or laid in shallow trenches, covered with soil and planted over. This ability to draw and accumulate nutrients identifies comfrey as a highly efficient "nutrient pump."

It is best to cut comfrey back as soon as it has flowered. To harvest before flowering might be better for nutrient content but I wouldn't want to disappoint the bees. It's not an unattractive plant when it's growing but once it has flowered it flops over and turns into an ugly blackened mess. When cut down it will sprout right back up with a smiley *"Hi there, I'm back!"* attitude, which used to drive me crazy. Certainly I would advise anyone who is planning to introduce comfrey to their garden to pick the spot carefully, in an isolated place that won't be needed for anything else, ever. After one initial planting my comfrey seemed to colonize several areas quite rapidly but since then has settled into a holding pattern.

It is my experience that comfrey is virtually indestructible and impossible to eradicate. I was quite intrigued recently to read an account of successfully sheet mulching with cardboard over a comfrey patch to create a comfrey free veggie plot. I can't imagine what kind of comfrey that was! It seems to me that comfrey is indestructible. A couple of years ago we inadvertently stacked a sizeable mound of fence posts on top of a comfrey patch. We'd forgotten it was there until it started poking shoots up through the top of the four foot high pile of posts. That's called determination!

According to some accounts it also makes good feed for poultry and people. However there are conflicting schools of thought on this and, at the first mention of possible liver and kidney complications, I decided that there are enough edible greens to pick from already and that comfrey can remain solely as a compost crop. One final word; there are several kinds of comfrey, the Russian variety seeming to be most preferred, perhaps because it's a hybrid and therefore sterile and unable to reseed itself. This has to be a plus in my mind as I shudder to think of a prolific windblown seeding in my garden. It's easy enough to propagate by splitting plants, but any gifting of comfrey plants should be accompanied by a strong warning: they are aggressive and once established, seemingly indestructible. All that said, as dynamic accumulators and natural nutrient pumps they definitely deserve a place in any permaculture plan.

One of the most positive steps towards improving soil might also be one of the easiest, as all that's required is a simple shift in perception—that is, to see soil as a living organism rather than simply as inanimate dirt. This promotes a deeper understanding of the complexities of soil, its likes and dislikes and its relationship to the plants it supports; and with this understanding comes a visceral connection between gardener and soil.

✳ Soil, a living organism, needs its own special "blankie" of organic mulch in the fall.

Not so long ago my gardening year always began with a thorough rototilling of the soil. My understanding was that the soil needed to be broken into fine particles so that delicate root systems could reach down, spread about doing their thing, sucking up all the manure I had so thoroughly mixed in with all my extended mechanical churning. What I didn't know was that the constantly rotating tines of the tiller were in fact doing more harm than good by destroying a seething, underground metropolis of symbiotic relationships, operating on a microscopic scale.

Without going into too much scientific detail (because in truth I'd be in well above my head if I did) it is these colonies of beneficial microbes and bacteria that are in fact responsible for ensuring that nutrients are readily available for the tender roots. This process happens only after the worms, beetles and other insects have processed the nutrients in their own inimitable ways.

Because of their minuscule size it's tempting to underestimate the impact these little critters have, but this would be a big mistake. The difference healthy living organisms can make in garden soil is huge. Digging, never mind extensive tilling, destroys their communities and throws the whole finely balanced system into complete disarray. I have come to the understanding that, once a fairly healthy soil mix with a good percentage of organic matter has been established, any form of digging is destructive rather than productive.

I said the mental shift was easy but in truth it isn't *that* easy. Breaking with long held traditions can in fact be quite difficult. My big problem with the no-till (no-dig) method was coming to an understanding that compost and manure layered on top of the ground can benefit the roots and capillaries several inches below. It took a while for me to understand how worms start the process by feeding up top and casting below and how the feeding chain continues through smaller and smaller insects right down to the microscopic miracle workers previously mentioned. My other problem was that it seemed just a little too easy—but it is true, the no-dig method not only produces far better results, it is also much easier on the back!

A couple of other things to keep in mind: soil, as a living organism, doesn't like to be stood upon. Anything that compacts the soil also flattens the air pockets and therefore deprives the microbes of oxygen. Even more damaging is compacting soil when it is wet. This not only destroys air pockets, it also further compromises any ability the soil might have (depending on the amount of organic material present) to spring back into shape. It's useful to remember that bricks are made by compacting wet clay.

Like many living organisms, soil needs time to renew. During the growing season the soil "engine" is actually running at its slowest, even as the growth cycle above reaches its peak. When the harvest is finally over for another year, the soil needs to be lovingly tucked in under a generous blanket of straw or other organic mulch. Around here kelp and eel grass are always most available

✱ Beneath its frozen crust the frenetic activity of jillions of microbes is renewing the soil's fertility.

in the fall, so this is when I harvest my couple of truck loads off the beach and spread it directly on the garden beds. I will also add what manure and leaves are available and then cover with straw. Putting the soil to bed in this way feels like an act of appreciation for the abundance it has provided and a way of honouring our natural world (another of the founding principles of permaculture). More importantly, perhaps, it creates optimum conditions for repair and renewal. Looking out at a garden, blanketed in snow and held tight in the clutch of the Ice Queen, it's hard to imagine the frenetic activity deep below, as nematodes, protozoa and all their microscopic relatives prepare a smorgasbord of nutrients fit to provide us with next year's crop and the finest of food.

To Dig or Not to Dig

For years I planted regular "in ground" veggie plots, carefully laid out in traditional rows. I had seen gardens and knew what they looked like, sort of. One of the (many) mistakes I used to make was mounding every row (usually way too high) because I thought that's how gardens were planted. I did mulch heavily between rows which definitely helped to keep the moisture in around the roots but, even so, during hot dry spells I'm sure that many of the plants had difficulty keeping their roots cool and moist. In retrospect, I believe my misconception arose partly from the fact that some plants, such as leeks and potatoes, like to be mounded up, but only after they already have a vigorous root system established.

Another problem I found with a traditional veggie plot was that, no matter how much organic material I added, the soil would compact and seemed intent on reverting back to its original clay-hard state. As already stated, I'd rototill like crazy every spring until I was knee deep in a sea of fluffy rich soil but then I'd stomp it all down as I worked planting the rows. To further compound the problem (and the soil) I'd work the soil when it was too damp. This is an absolute no, NO! Remember how bricks are made.

I have since reworked most of my plots into raised beds and I would recommend this method absolutely, if only to avoid putting any heavy pressure on the soil. Raised beds can be as long as you like but not more than four to four and a half feet wide, only as wide as will allow an easy reach to the center of the plot from either side. One of the prime rules of raised-bed gardening is that the soil is never stood upon. This ensures that it doesn't get compacted and doesn't need tilling or deep-digging. It also makes it much easier to control the quality of the soil.

With all types of soil it is essential to replace the nutrients the plants feed on, and here there is a grey zone (in my mind at least) as to how much the manure/compost etc. needs to be actually dug into the ground. There's no doubt the nutrients must be added but opinions on how much the soil should be disturbed vary. One opinion is that all the additions should be simply layered on top of the ground allowing worms and other organism to do the mixing. This could be classified as the "no-dig" method. As already mentioned, the main argument against digging (other than it is hard work) is that any strenuous disruption of the soil destroys

✳ Moving aside the winter blanket of organic mulch reveals weed free, friable soil ready for planting.

the guilds or working relationships that the microorganisms have established, and those relationships are essential to healthy plant growth.

Seeing soil as a living entity rather than as an inert substance makes the concept of nurturing the soil as well as the plants much easier to understand, and also avoids the risk of doing more harm than good through excessive "soil preparation." The no-till method requires that the soil be heavily mulched over winter using organic materials such as leaves or straw. This prevents the proliferation of weeds as well as the leaching out of nutrients. At planting time the mulch is simply moved aside enough to allow the seeds or transplants to be introduced to the soil. Even if the no-till method is not being used, it's always good to mulch over winter, as it still discourages weed growth and prevents nutrient loss. Also, turning the soil over will be much easier if it hasn't been hammered hard by ice pellets and hail.

Another approach, midway between no-till and turning the soil in the spring, is to scratch the top couple of inches with a rake or a hoe, gently mixing the added nutrients into the top layer of soil and creating what is referred to as a "tilth." This is the method I tend to favour. Either way certainly beats struggling with a noisy, cantankerous gas-powered tiller. I would never do that again to an existing garden and consider power tilling to be a questionable extreme, even when setting up new beds. It is certainly not viable as a yearly maintenance.

Slugs were always my nemesis and with raised beds I have prevented their access considerably by stapling copper mesh along the top sides of the walls that surround each bed. Copper gives slugs an electric shock and the wire therefore acts just like an electric fence. Fifty or one hundred foot rolls of this wonderful mesh, which looks a lot like an uncoiled pot scrubber, might seem fairly pricey at first but they can be used year after year and, compared to the cost of lost plants, not to mention crumpled dreams, they really are worth every penny. And while on the subject of pennies, they can also be circled around the base of tender plants to create a no-go zone for slugs. Since I haven't found this particularly successful, I suspect it might be one of those high appeal, low delivery concepts.

Another advantage of raised beds is that they can be constructed over inhospitable ground. I have read accounts of raised-bed gardens producing well even when they are constructed over concrete and asphalt. To do this I imagine they would need to be at least twelve inches deep and even at that might not be favourable for deep rooting crops.

Raised beds are not difficult to construct. They can be as simple as earth mounded up to form beds but I prefer some kind of wall or edging to hold the soil in place. These containment structures can be made of bricks or wood or even bales of hay. My raised beds are constructed on what used to be my regular garden plots so presumably fairly good soil underlies them for a foot or so and walls eight inches high are quite sufficient. If I was building over rocky clay or concrete I would want the walls a few inches higher. My walls are constructed out of two by eight planks and my longest

✳ Raised bed constructed to fit contour lines.

bed is twelve to fifteen feet long and about four and a half feet wide. As mentioned, it is the width (typically four to four and one half feet) not the length of the bed that is crucial but every foot of length requires two extra steps to reach the corresponding position on the opposite side. Beds any more than fifteen to twenty feet in length would be a mistake, in my opinion.

The other width that requires careful consideration is that of the pathways between the beds. They need to be wider than might seem necessary, ideally three but definitely no less than two and a half feet wide. It is easy to look at dormant boxes of soil and believe that pathways eighteen inches wide between them would be plenty wide enough, but wait until those boxes are overflowing with abundant growth and then try crouching between them to harvest without breaking off exuberant leaves and branches. At first it seemed to me that I was losing valuable planting area to the pathways but, because raised beds are typically more fertile, they can be more densely planted and therefore produce a much

✷ Raised beds with early spring growth—the pathways appear to be adequately wide.

higher yield per square foot. Any perceived loss actually translates as gain and the extra few inches of width along the pathways will definitely be appreciated, guaranteed.

One truly innovative permaculture idea, designed to maximize available ground and minimize difficulty accessing the plants, is the keyhole garden. It uses a meandering perimeter, much like the shape of a keyhole, with the pathway surrounding the outer perim-

✳ Midsummer raised beds gone "crazy"—the pathways disappear.

eter but also leading inwards to create an open, central space where the "key" or, in this case, the gardener can go. This design has the major advantage of allowing easy access to a maximum amount of space. This works well if the garden is more or less at ground level but I would hate to have to construct a raised bed wall around a keyhole garden. I have seen such a garden edged with rocks but my experience with rocks as containment walls has always been negative. I found that sooner or later weeds infiltrated the cracks between the rocks and wove their roots so densely around and under the rocks that it became impossible to eradicate them. This is why I personally prefer straight planks.

Straw bales can be used in a couple of ways to "construct" a garden, depending on the availability of space and bales. They can be arranged to enclose a square or oblong inner area which, when filled with suitable soil mixture will be become a raised bed garden. I would probably not choose this method if there was another option because I believe the wet straw will, over time, create a perfect breeding ground for slugs... but then slugs are my obsession.

✱ Newly constructed straw bale bed.

✴ Straw bale bed after a couple of years.

The other way to use straw bales, and this is especially good on particularly infertile strips and nooks, is to plant directly into the bale by hollowing out several holes to a depth of six or so inches, and of a similar width. Fill these with rich soil or compost and plant directly into them.

The "lasagna" garden is another form of raised bed which derives its name from the layering of various organic materials. It can be built directly on the poorest of soils, with a layer of wet newsprint and cardboard laid first to mulch out any potential weeds. This base is topped with whatever organic materials are available, such as grass clippings, spoiled hay, leaves, manure and compost, interspersed with layers of soil and perhaps some limestone. (The pelletized lime is assimilated more rapidly than regular powdered stone, but dolomite is perhaps the best of all.)

It's best to spread the "ingredients" of this lasagna garden in several thin layers. For example, instead of one 2" layer of compost, two 1" or better still four ½" layers will work best, in that it's less distance for soil organisms to travel to get what they need and the mix will therefore become really good soil more quickly. There is no particular list of ingredients for this recipe but variety is the key. Just about anything organic goes, as long as the compost is covered by a layer of soil and then a final top layer of straw to mulch. Lasagna beds can be planted as soon as they are built but it is in their second and subsequent years, after the ingredients are well into the decomposition process and the worms have had a chance to mix things up, that the rich, fluffy soil will really start to bloom.

Yet another way to create fertile raised beds is the hugelkultur method. This technique is, I believe, of Austro-Germanic origin. It mimics the natural woodland fertility cycle, which also provided the inspiration for Analog Forestry, a system quite similar to permaculture in many ways. As with the lasagna beds, hugelkultur employs a method of layering, but in this case most of the organic material is woody detritus. Wet newsprint and cardboard are layered down first then topped with small logs, layers of branches and twigs, starting with the coarsest and ending with leaf mould, grass clippings and straw, all of which is then topped with soil. The heat produced by the decomposing wood promotes early vigorous growth of plants such as beans, cucumbers and other squash, all of which like their roots in warm soil. These beds are great for early spring plantings but we also like them because they provide a use for all the storm debris we are still clearing away. They also provide a workable solution for uneven, rocky ground with little or no naturally occurring topsoil.

The productivity from such beds is really astounding and seems to increase as the woody materials decompose. Perhaps it shouldn't come as such a surprise, considering that forests produce giant trees using just this method, but the bumper crops that come tumbling out of a relatively small hugel bed we started several years ago are indeed amazing. This is definitely my preferred system of alternative garden bed and it surprises me that it doesn't seem to be as popular as it deserves to be.

* Rotting logs are layered down first.

* Smaller logs are next.

* Finer branches are laid on smaller logs.

* Leaves, straw, humus, compost and soil form the final layers.

* Prolific growth in a first year hugel.

I have also learned that rotting logs in trenches, either as a base for hugel beds, or simply covered with soil, create equally warm, fertile beds, with the added advantage that the rotting wood helps to conserve moisture as it acts like a sponge, holding large reserves of water to help the plants through dry times. This is a perfect example of how to maximize on a naturally occurring process: the rotting wood creates a haven for microbes and bacteria to go crazy in and their activity transforms this excess of nutrients, making it readily accessible for the plants to feed on. Permaculture encourages the careful observation and use of natural processes such as this. In nature nothing is wasted. Natural systems tend be cyclical in their own right while at the same time creating symbiotic relationships with other closed but neighbouring systems.

These kinds of inter-relationships are highly complex and it's easy to understand how the smallest shift or imbalance can have huge consequences. At the same time, respecting these same systems can lead us to a self-regulating and sustainable existence, which produces maximum yields for minimum expenditure. It's important to think of expenditure in terms of energy consumed, carbon output and global impact rather than simply in terms of dollars.

From a permaculture point of view there are no (insurmountable) problems, just catalysts for creative solutions. My "Herbal Hugel Heap" is a good example of this. The herb spiral is in some ways a quintessential permaculture icon in that it uses a minimum amount of footprint area by curving upwards rather than sprawling outwards. It also creates a variety of microclimates (in a very small area) which in theory will cater to the specific needs of a variety of herbs: top plants have dryer soil and more sun, plants at the base of the west side avoid the noon day sun and enjoy more moisture, etc.

My "Herbal Hugel Heap" was constructed in response to a problem rather than a need to create such a variety of growing conditions. The problem was twofold. It consisted of several large partially uprooted tree stumps and an excess of well-manured duck bedding. Not only were the uprooted tree stumps an unpleasant reminder of the carnage left in the wake of the hurricane, they also provided a mustering point for several kinds of weeds that we just didn't want spreading into the adjoining vegetable beds. This problem area was by the gate leading into the duck compound, conveniently close to the manured hay that needed to be moved.

First I mulched with newsprint and cardboard, as best as I could, over and around the stumps to form an irregular cone shape. Next step was to begin layering the old hay bedding on top to construct a conical mound. By walking up and down the mound in a circular fashion I was able to create a fair facsimile of a spiral. To my great good fortune I happened across a variety of herbs at end of season, give-away prices which I was thrilled to bring home and re-pot. I didn't feel comfortable putting the plants straight into the duck bedding for fear of it being too "hot," so instead I sank the pots into the mound, along the spiral path I had shaped. The pots were a good size providing plenty of growing room for the herbs and also creating stability and erosion control on the mound.

✳ Immovable stumps mulched first with cardboard.

✳ Topped with spoiled bedding from duckshed.

✳ Pots add structural stability.

✳ Herbal hugel four years later.

A troublesome eyesore has become a pleasing focal point where I stop each morning for some chocolate mint to nibble on while gathering eggs and feeding the ducks. Will all the herbs I planted thrive equally well? I doubt it but that's okay. Some of the thyme, the marjoram and mint plants appear to be thriving, as do the garlic chives and camomile. Any spaces can easily be filled with some of the annual herbs started indoors each spring, such as basil and dill. I'm really pleased with my Herbal Hugel Heap and feel it has proven to be a very happy, low-cost solution.

The fact is that food can be grown in an old boot if it's well tended (the plant, that is, not so much the boot). With a little imagination just about any place will start to look like potential growing space and so it's good to consider several other factors, which are just about as important as the soil itself. This is where permaculture planning makes such a positive difference and yet it's an area that might initially make permaculture seem more complex than it actually is. For example, the terms "zones" and "sectors," because they are not typical household words, might at first seem a tad intimidating.

＊ A simple, "modified" keyhole bed.

In truth, all these styles or methods of setting up a successful garden bed are best thought of as general concepts, rather than as inflexible rules and recipes. The main function of a keyhole garden, for instance, is to extend perimeter edge while also creating easy access to all areas of the bed. Some diagrams will show keyhole gardens with edges that are positively "frilly" and in my experience that's just not feasible. However, I love the concept. Our first experiment with hugelkultur was constructed over a circular plot that had been invaded by couch grass or twitch grass (*Elymus repens*). To quote The Royal Horticultural Society: "It's an old enemy for many gardeners. Its wiry, underground stems and creeping shoots pop up around garden plants and before long can take over a bed. As a perennial weed thorough killing or eradication of the roots is necessary." Couldn't have said it better myself! Its tenacity is really quite admirable if it wasn't so despicable. Anyway, the hugel bed we constructed over it did the trick. This particular scourge was finally eradicated. The bed is roughly circular, amazingly productive and approximately eight feet in diameter. Harvesting around the perimeter was easy, reaching the center crops, impossible without stomping on the outer ones. Using the concept of a keyhole garden, four wooden platforms, strategically placed, transformed this bed into a hybrid keyhole/mandala garden. Think a four leafed clover, with a central hub. One of the many wonderful things about permaculture is that it provides the tools to encourage flexible, creative problem solving. Have to love that!

The consideration of zones, simply put, ensures that areas which require daily or regular visits, such as the salad garden, are situated close to the house, whereas areas which require little or no attention, such as areas left "wild" to encourage natural habitation, are situated furthest away. It is simple common sense, as are most of the permaculture rules, simple common sense clarified and articulated. The process of thinking in terms of sectors defines the physical attributes of a place, such as the path of the sun and the prevailing winds, possible sources of flooding or pollution, and other elements which will influence the positioning of gardens and outbuildings. There's much more that can be said about zones and sectors, but in a later chapter. Right now, it's time to be putting a For Sale sign on the rototiller, and delegating the spade to that most distant, cobwebby corner of the garden shed, in behind the hammock and the lawn chairs.

Green Thumbing It 101

Even though I've always liked vegetables and been ready to taste, and usually enjoy, anything new the produce manager might put out for sale, when it came to growing my own I was astoundingly conservative. I couldn't see beyond the traditional suburban crop picks of tomatoes, cucumbers, lettuce, radishes and beans. I also thought everything went into the ground about the same time, around May 24th, and didn't require much attention after that until it was harvest time. I'd love to have a convincing excuse for such limited and unimaginative thinking but it simply was what it was.

In comparison, I now plant about thirty to thirty-five different crops, all of which are basic, run of the mill vegetables, nothing exotic or in the least difficult to grow. I begin planting seeds in March and will still be sticking stuff in the ground in September. This massive shift came about gradually as I began to learn more about the individual preferences of plants. Plants, not at all unlike humans, have very individual likes and dislikes. Sure, we might endure less than optimum conditions, but we probably won't thrive in them. It's just the same with plants. Peas like to be planted in trenches while squash prefers mounds, for example, and peas flourish in temperatures that squash would refuse to tolerate.

In areas where the spring comes late, and summer's heat is slow to arrive, squash might be best started indoors. The reason I say "might" is because squash plants can be a little difficult to transplant and can be shocked into a "no-grow" state by the sudden transition from one environment to another. This is where the important step of "hardening off" comes in.

Even after they have been hardened off young squash plants will still go into a major sulk if their roots are disturbed so it's important to slide the clump of potting medium out of the pot, carefully transferring the plant into a waiting, well watered hole in the manure mound that is to be its permanent home. I like to hedge my bets by seeding the mounds as well. So that the in-ground seedlings don't get disturbed by the new arrivals I dig the transplant holes in advance, keeping them open with an empty can or yogurt container. This might sound complicated, but it really isn't and it does provide a certain insurance that either the transplants or the seeds will grow into healthy plants. If, as in a perfect world, both plantings are totally successful, and the mound is simply bristling with eager little plants, the grim reaper (gardener with scissors) must be called in to thin the mounds down to two or three of the healthiest plants per mound by snipping off, *not pulling out*, the excess. Pulling out the excess will disturb the surrounding

Any plants, but especially plants that are sensitive to change, that have been raised in a controlled climate (greenhouse or grow light) need to be introduced gradually to the fluctuating temperatures and air currents they will experience in the great outdoors: on the first day they can be put outside for a few hours in a shady, sheltered spot and the next day for a while longer, with this step repeated for several days. An overnighter comes next but only when the nighttime temperature differential isn't too extreme, and then finally some exposure to full sun before sensitive plants are "hardened off" enough to be planted in the ground.

✻ There's a fine line between densely planted and overcrowded. This bed is overcrowded, partly due to a few infiltrators that snuck in uninvited.

root systems of plants being left to grow and it also robs the soil of growth into which it has already invested precious nutrients. Left in place these excess roots will rot, giving nutrients back to the soil and also leaving minuscule channels to aid aeration and the permeation of water.

This process will seem especially wasteful to anyone who has paid a couple of dollars for the scant handful of seed that generated the overpopulated mounds. Overcrowding is even more wasteful! None of the plants will be able to flourish and will eventually choke each other out. Someone told me years ago that I wasn't "brutal" enough to be a good gardener and it was true at that time. Even now it really bothers me to destroy little plants but, after learning the hard way, I know that at times the head must overrule the heart.

* When cilantro goes to seed it becomes coriander which is an easy seed to save for next year's cilantro crop. It's provides several important health benefits and is used in numerous recipes, especially those with Indian or North African influence.

Back to the scant handful of seeds (eight or ten) that averaged out at about twenty-five cents apiece. Seed saving and seed sharing is an essential part of any garden and especially of any perma- culture plan. It is so easy to save seed, not to mention money,

✳ Chamomile flowers produce viable seed; they make wonderful tea and are also delightful to work around, as their aroma perfumes the air.

which can easily run into three figure amounts for a typical basic selection of vegetables, with a few floral picks thrown in to attract pollinators.

Many vegetables produce quite attractive flower spikes which then go on to produce seed. Arugula and onions are two diverse but equally appealing examples. Prior to forming its seed arugula puts on a pleasing display of delicate, cream colored flowers which are happy to provide the "wallpaper" in a mixed bouquet, whereas onions send up tall flower spikes topped with large balls formed by hundreds of tiny purple or white blossoms. They demand attention!

Some "vegetables" produce their seed as part of the flesh, which technically makes them a fruit. Tomatoes are a perfect example. Squash is yet another fruit that might be seen to be masquerading as a vegetable, but this does not alter the fact that all those giant seeds, that get scraped out and often thrown away, can

* This massive tower of vegetation came from a parsnip that overwintered. It produced a huge amount of seed that looks a lot like
dill seed.

be dried, saved and grown as next year's crop or roasted and eaten
as a highly nutritious snack. It really is that simple and is therefore
somewhat mind-boggling that commercial seed companies have
thrived as well as they have. Seed saving and swapping is the most
natural, affordable and interesting way to ensure the sustainabil-

✳ This summer squash will produce enough seed to grow innumerable plants next year, providing the seeds are dried and stored correctly.

ity of any garden. It also builds networks with kindred spirits and thus fulfills another principle of permaculture, which calls for the strengthening of community, the transfer of knowledge and the empowerment of others. More on seed saving in a subsequent chapter—for now I'm just planting the thought.

As earlier mentioned, this book is not meant to be the quintessential book on sustainable organic gardening. There are already several brilliant gardening books that fit that description (see Suggested Reading). This book is more of a primer, words to broaden perceptions, lay the ground work for further exploration and flag some of the more common pitfalls.

Having established that planting is an ongoing process and that most plants have their own peculiarities, it can be useful to divide a typical year into three time allotments: early, mid-season and late. How many people have planted peas and beans at the same time and wondered why only one thrived? Or had only one feed of crispy sweet lettuce before it all bolted and turned bitter? Proper timing is essential yet was something I never gave a second thought to years ago. Now, with more experience, I find it stings to see certain seeds offered for sale in June or July, considering they could not even come close to fulfilling the promise of the succu-

✳ This broccoli and lettuce is thriving in the cool temperatures of early spring, along with some Egyptian onions left over from the previous year.

lent harvest pictured on the packet. It saddens me to think how many wanna-be gardeners have tried once and never again because of seed displays that by their very presence hint at the impossible. It's just lucky that I'm stubborn!

I shouldn't be so hard on seed packets because for the most part they are quite explicit regarding the plant's preference: "as soon as the risk of frost has passed; not until the ground has warmed up; plant indoors three weeks before the last frost" and so on. I guess it's not their fault if some people (like me) don't read instructions carefully enough or think they are just there to fill in the empty space on the back of the packet. It's important to note that *as soon as risk of frost has passed* does not necessarily mean the middle of May. Spinach and kale are especially hardy and thrive in the cool, and even quite cold, early days of spring. Sometimes the gardener (once again read me) who doesn't much enjoy bundling up against a brisk wind to work the cold damp soil, simply prefers to think this can only mean mid-May. Many seeds are happier going in much earlier than that, and in fact some seeds refuse to germinate in warmer temperatures and even when induced to sprout will begin going to seed as soon as the hot weather begins.

Working damp soil is an absolute no, no! When soil, whether cold or warm, is worked damp it clumps up and forms, in essence, bricks or clods that constrict the delicate roots as they reach out in search of nutrients. As the soil is compacted tiny air pockets are squeezed out of it. This not only deprives the plants of oxygen, these spaces would have also enabled water to trickle down to roots thirsty in the summer's heat. I know I've said that once already but the point deserves repetition because in the first heady days of spring it is so easy to be tempted to dig in the dirt.

The temptation is most easily avoided by preparing the soil in the fall and covering it with a straw or leaf mulch. If the soil is damp, as it very well might be in spring, it is not necessary to touch the surface at all. The lettuce, spinach and kale seed can be sprinkled on top of the ground and covered with a thin blanket of commercial potting mix. If a hard frost is forecast after the seeds have stuck their green shoots out into the world, the little ones will probably be fine if covered with a generous mat of straw.

In this chapter we have just barely touched on the rhythms and intricacies of Mother Earth. As we explore further it will surely become more and more apparent that we live in a perfectly balanced universe, alive with invisible rhythms and flows that influence

A word of warning: Beware of slugs! The first green shoots of the season taste just as good to them as they do to us and they seem to be very good at hiding out in straw mulch. Best to let the ducks free range on the mulch for a few days before seeding. Why ducks? Well, it would appear that slugs are beyond gourmet to the refined gastronomy of a duck. They will sweep any garden bed with diligent, energetic precision until they are certain that every last sublime morsel has been found, fought over and consumed. Also, ducks will be going crazy with mating fever at this time of year and it might just take a little pressure off the ladies if you send the drakes on a slug hunt. No ducks yet, you say? Well, crushed clam or egg shells, or wood ash, sprinkled liberally around the perimeter of the planting area right on the soil, close up to the emerging seedlings, would be my second choice and also my backup plan. When it comes to slugs I don't take any chances.

Another plant that I have noticed attracts slugs is the humble and unfairly maligned dandelion. Slugs will cluster in tight around the base of the dandelion stems and are easy to pick off and dispose of (i.e., feed to the ducks). Dandelion leaves are wonderful added to an early salad and they are a treat par excellence for the rabbits. You don't have rabbits? Oh, but you will! (Insert maniacal laughter here.) Dandelion wine and dandelion beer are two other more celebratory uses for this pretty yellow flower that just cannot understand why it evokes such disdain, especially from "Townies."

✳ The young ducks are preparing to wage war on slugs.

every living thing. The cycles of the planets, however distant they may seem, exert an influence on each tiny seedling. Such ancient beliefs are still adhered to by many who believe root crops are best planted on the waning moon while plants that are harvested above ground should be planted shortly after the new moon. ("What's in the ground, plant in the empty of the moon; what's above, in the rising of the moon"—an old saying collected from both English and German farmers in Leminster, Nova Scotia, by Dr. Helen Creighton, renowned collector of Maritime Folklore.)

Certainly the "Alder Cutters' Moon" was one of the things that helped me decide that my husband was indeed the man of my dreams; he was the only person I'd ever met who had heard of such a thing and, better yet, was able to explain why alder bushes aren't likely to grow back when cut around the full moon in August. The carbohydrates (as storehouse of energy) are up in the leaves and stems when the trees are cut and the roots don't have time to re-stock nutrients for regeneration before a killing frost comes and puts an end to them.

Studying the natural world enables us to more fully appreciate the astounding and intricate beauty that surround us. The perma-culture way helps us define our place in this amazing design. The more carefully we observe this design, the more successfully we will be able to integrate with it.

Easy Starter Crops

I've chosen to mention the following four vegetables here because they are all relatively easy to grow. Also, each has a preference for when it is planted:

Garlic, for instance, likes best to be planted in the late fall. It might have taken all summer to prepare soil and construct a new bed and it's satisfying to be able to plant right there and then, rather than have the bed remain empty until the next spring. Garlic is the perfect choice.

Chard likes to be planted in early spring. It is very hardy and will grow reasonably well in mediocre soil.

Potatoes are planted mid to late spring and don't require the kind of perfect soil that develops after several years in a well-tended plot. They also don't require a lot of soil as they can be heavily mulched with organic material to provide all the cover they need. They're great for starter plots, especially as the mulch rots down and increases the amount of organic material in the existing soil, thereby increasing the actual amount of soil as the bed is used.

Finally, squash can be planted considerably later, in early summer, in an easily constructed mound, as opposed to a well worked bed. They're good for impatient gardeners (and aren't we all?) and are usually quite prolific, pleasing in both taste and performance.

* A clove of garlic going in the ground. Note the placing of the cloves already laid in place. A little close perhaps but they did fine.

Back to garlic—it's a very satisfying and purposeful crop to plant. I say purposeful because it almost seems like it does have a purpose as a curative and a preventative, in addition to being a surely most necessary ingredient in so many dishes. In medieval times it was also thought to keep the devil at bay and, worn around the neck, it was believed to protect against the plague. While I won't argue against any of these uses, I would like to add my theory to the list. I believe garlic, by way of its rather specific needs, demonstrates what a simple thing it is to obey cycles and seriously consider planting times. It also strongly encourages crop rotation and late season planting.

Years ago, when I didn't have much of a clue about gardening, I tried planting garlic. All I harvested were some brittle stalks with slightly nubby ends. I decided garlic must be very difficult to grow and didn't try again. I had planted the cloves in June (along with everything else) and harvested in September. If I had waited until the next spring to harvest I might have had a great crop. Garlic has

✷ Fall planted garlic, the following July.

very specific ideas about when it needs to be planted. It likes to be planted in the late fall for harvest the following summer. It's as simple as that. Ideally the cloves should go into the ground about three weeks before the first frost. Early enough that it can put roots down but not allowing enough time for substantial shoots to emerge above ground.

The fall planted garlic will sprout in early spring and grow slender leaves looking not unlike daffodils at first. Sometime in early to mid summer they will produce scapes. Scapes are thin but rigid stalks each topped with a pointed green cap which, if left, will open, flower and produce bulbils. Before this flower appears the scape will curl around on itself creating a circle about three inches in diameter. These scapes look very elegant at this stage, a wonderful addition to any flower arrangement. More importantly, they make delicious pesto.

Several weeks after the scapes appear some of the outer leaves of the plant will start to turn brown and wither. This means it's

✳ Note the graceful curl of the garlic scapes.

time to harvest the garlic bulb. If left too long past this stage the outer, papery wrap of the bulb will split as each clove begins to swell prior to sprouting. The cloves don't store as well in this condition and may taste a little bitter.

There are three or four main types of garlic. Some have larger but fewer cloves than others and the intensity of flavour will also vary. The larger cloves are not necessarily the most pungent, but

✳ Harvested scapes waiting to be transformed into pesto.

The exact time to harvest garlic cannot be directed by a specific calendar date, as it can vary and is doubtless influenced by growing conditions. It's necessary to read the progress of the plant by carefully observing any external changes. The scapes, or flower stalks, are clear indicators of this progression. When the scapes first appear they stand upright and then, after a week or so, the pointy tips will begin to droop quite elegantly. This is when the scapes are best harvested for pesto and stir-fries. If left they will continue to curl in on themselves and will eventually form circles approximately three to four inches in diameter. The garlic bulb itself, still hidden beneath the soil, is not ready to harvest until several weeks after the scapes have begun to curl. At this time the tips of some of the outer leaves of the plant will begin to turn brown and wither. This is when to find a sturdy digging fork and begin to harvest.

This "reading" of a plant's readiness by careful observation of its condition is akin to phenology (the formal study of the life-cycles of plants and animals) in that it determines the cycles of a plant by its behavior rather than by the flip of numbered days on a calendar.

✳ Garlic harvest hung to dry in the garden shed.

they are certainly easier to skin. It is always a good idea to plant more than one variety, not only to discover which type you prefer but also to ascertain which type of garlic most likes the soil and climate conditions you have to offer.

Each clove planted will produce a full bulb. Bearing in mind that a braid of homegrown organic garlic makes a pretty phenomenal gift, it is better to plant more than you think you will use. Each bulb will need to be lifted with a digging fork as they will have developed a clump of short but tenacious roots. These roots can be trimmed away but the leaves are left to be tied or braided and the garlic is then hung in a cool dry place, such as the garden shed, to dry. One disastrous winter I left our whole garlic crop hanging in the shed too long and it froze. Not good!

N.B. Although the optimum planting time for garlic is in the late fall, it is possible to plant in the very early spring as soon as the frost has left the ground and still have a relatively good yield by harvesting much later in the year.

Potatoes are another simple, yet highly satisfying, crop to plant. Having said that, I should add that when it comes to specifics they are one of the few crops Calum and I disagree on. I believe that potatoes can be grown in a deep bed consisting of nothing but eel grass (seaweed) whereas Calum insists that this is not possible as they require at least a little soil to produce fruit. This creates a dilemma. Certainly, I don't want to be responsible for any disappointments come harvest time, but having known a couple of old timers who survived on not much more than potatoes grown in eel grass (along with as much salt fish as anyone might want to eat in a lifetime) I like to plant in eel grass. A similar effect can be achieved by laying potatoes on the ground and layering with straw as the shoots emerge.

Why buy seed potatoes when some of last season's spuds have sprouted in the basement? Seed potatoes are grown specifically to produce many offspring. They are planted late and harvested early so that most of their growth capacity remains untapped. This untapped potential allows them to flourish with vigorous growth when given the opportunity. Last year's wizened potatoes by comparison have already used up much of their growth potential by growing to their full capacity and will not have multitudinous progeny. It is possible to produce one's own seed potatoes but even this is not the best idea. The most common diseases that affect potatoes, such as scab and center rot, tend to accumulate over time. Using seed from a single source allows initially slight problems to compound with each planting.

Similarly, planting in a designated "potato bed" year after year is just asking for major blight and rot issues. Crop rotation is essential to healthy gardens and it applies to most if not all vegetables. Certainly it will be mentioned again but as potatoes are particularly susceptible to accumulative cycles of disease, now is as good a time as any to introduce the concept—not that there's anything new about crop rotation. I vaguely remember being told it was one of the key points of the Agrarian Revolution in Medieval Europe but somehow, back in grade school history class,

✳ Potato bed in bloom. Commercial potatoes growers even spray to
kill the blossom (and therefore speed up fruiting) but what a shame
to destroy these attractive purple and yellow blossoms before
their time?

that seemed like such a totally irrelevant piece of information.
Who could know?

There are many different types of potatoes, with almost as
many differing attributes. Broadly speaking they fall into three
main categories: early, mid and late season. Some store much bet-

ter than others, some are dry and therefore mash well, while others are more moist and perfect for baking. Some have yellow flesh, some have red skins, some have smooth thin white skins that don't need to be peeled, while others can be deep purple and look somewhat more unusual. It's a good idea to plant more than one kind of potato, perhaps an early, a mid and a late variety, taking care to mark which is which. You might think you will remember which is which come harvest time but chances are pretty good that you won't. It has been my experience that the earlier varieties tend to have the more tender skins while the later ones have thicker skins which protect the potato better during prolonged storage.

Potatoes are tough, salt of the earth types as far as personalities go and very easy to grow, but they do not like the cold. Planting too early will result in ruined seed and harvesting too late, after a frost or two, will result in a slimy mess and much disappointment. Rule of thumb: if you think it could be time to plant potatoes, you might want to wait a week or so. Premature planting in soil that

✳ Newly sprouted potatoes in a "potato box" which is moved each year to ensure proper rotation. Note the comfrey peeking over the edge, waiting to get chopped and mulched around the potatoes.

has not had quite enough time to warm up and there is still risk of a late season cold spell, will often result in the need to replant. And that's no fun. This might sound like a contradiction but a later planting can often result in an earlier harvest. Plants that don't like cold soil are easily set back and are then slow to recover.

Potatoes are in fact a vine and can be encouraged to grow vertically by adding mulch every week or so as the head of the plant appears on the surface, poking up towards the sky. One way to do this is to plant the seed potatoes in a "box." Ours is about five feet square and effectively a "coral" as it has no base. The potatoes are placed directly on top of the soil and when they are showing sturdy growth of four to five inches they are covered with more soil, straw or seaweed. This process is repeated until the box is full. It's best if the box is constructed of stacking board-width squares so that at harvest time the layers can be removed to allow easy access to layer upon layer of potatoes.

This system is another demonstration of the less-for-more permaculture ideal as it uses minimum space for maximum yield. Any suitably sized horizontal container can be used in this way, as can a cylinder of snow or chain link fencing. Some people use stacked car tires which are perfect in shape and volume for a single potato plant but our concern over what the potatoes might be absorbing from this petroleum based product keeps us away from doing this. This reminds me to reiterate the importance of not using pressure-treated wood for anything in the garden. Treated wood might contain arsenic or other chemical compounds which can leach into the ground and be taken up by your food plants. The risk might be minimal but why take it?

Another rugged, easy to grow staple around these parts is Swiss chard. It won't germinate once temperatures rise above 60°F (15°C) so it needs to be planted in the spring. It will withstand many hard frosts and often manages to overwinter, especially when well mulched. It might not be everyone's favorite when many other choices abound in the summer garden but cooked up, fresh picked in February, it tastes positively gourmet. I like to sim-

✳ Harvesting potatoes is as much fun as a treasure hunt and the perfect way to spend a warm fall day. Almost as satisfying as eating them!

mer it with a clove or two of garlic and dress with a mustard/mayo sauce. It is also a perfect replacement for spinach in Turkish lentil soup and many other recipes that call for spinach.

Traditionally chard came with a thick white stem similar to the rib of Bok Choy. This is still probably the hardiest variety but next in line would be the red stemmed or "strawberry" chard. This is my favorite. There is also a rainbow variety which includes yellow, orange and peppermint striped stems. This is a novelty, very

✳ This rainbow chard stayed productive tucked under a blanket of eel grass until the hard freeze came.

attractive and perhaps more delicate in taste and texture than the heritage strains, but it seems to be more difficult to propagate, more inclined to bolt and is definitely less tolerant to frost. It is a good idea to plant a couple of strains of chard, but not too much of it. It grows vigorously and unless you have several friends and neighbors who love chard, using it all can become challenging.

Fortunately it does freeze well when lightly blanched, so chopped and frozen it's a viable option in winter soups and dips.

It's very tempting to overplant just about everything. Come March or April the earth looks so desolate and tongues are beginning to yearn for the taste of crunchy fresh greens. How can less be more? Well, yes it can, especially when all available planting space is filled to capacity with early stuff and you still have several later varieties you want to try.

If all the spring planted chard does happen to bolt through the summer, a fall planting is recommended. If it is a particularly warm fall, such as we are often blessed with, it might be difficult to get the seeds to germinate. This is when I start seeds in trays in the basement, which always remains quite cool. In the spring the seeds are planted directly into the ground. They're better thinned as they start to grow because a healthy plant, which is shaped similar to a large romaine lettuce, can easily produce a dense growth ten to twelve inches in diameter. It is always best to snip off rather than uproot when thinning so that the remaining plants are undisturbed and the "decapitated" roots break down, creating channels which enable air and water to penetrate the soil. An additional advantage to this method of thinning is that the small leaves that have been culled are all ready to go in the salad with no muddy roots to deal with. Cutting beats pulling just about all ways.

The three previously mentioned vegetables all require to be planted fairly early in the growing season. The next two suggestions prefer to be planted later when things have warmed up considerably: squash and pumpkin. Here I'm using the word squash to refer to summer squash, such as zucchini, and pumpkin to refer to those big orange globes, which I think of as winter squash even though they are actually mid-way between the two. There are many varieties of both winter and summer squash. Generally speaking winter squash take longer to mature. They have hard skins and store well. Summer squash fruit up more quickly and are better picked when they are young, not yet fully matured. Both have a variety of uses in salads, soups, stir-fries and baked goods.

In a new garden, a suitable growing space can be prepared for either of these species fairly easily. They are heavy feeders which need room and like lots to drink but don't like to have their feet (roots) kept wet. As the instructions on seed packages usually state, mounds of well-rotted manure work best. Compost mixed with some soil works equally well, if not better. Manure, compost and soil; well that's the bees' knees and yes, bees love the big yellow flowers that both species produce prior to fruiting.

It doesn't take much effort to set up a fertile mound (two to three feet in diameter and eight to ten inches high) on a patch of uncultivated, infertile ground. To keep down surrounding weeds I simply mulch with newsprint and cardboard covered with straw. The squash plants, which are vines, will need space because depending what they are feeding on (what is in the mound) they might well travel twenty or thirty feet. They can be trellised and trained to grow upwards, and they do climb trees. Fun! Unimpeded sunlight and regular watering are the two other necessities.

✳ Here the previously uncultivated ground was simply mowed, to accommodate two hastily constructed raised beds.

✱ Dog was not much impressed by the harvest they produced, but we were.

Occasional watering with manure tea, or fish fertilizers, makes a huge difference and in fact growers of humongous prize-winning pumpkins usually have their own secret recipes, which they feed to their pet plants on a daily basis in order to encourage their impressive girth. I believe regular consumption of chocolate cheese cake works in a similar way on some of us.

Zucchini and other summer squash don't grow on vines like winter squash, even though their requirements for healthy growth are quite similar. They look almost identical to their winter cousins during the first stages of growth but they don't travel. The plants shape up to look more like wide, low growing bushes than vines: two to three feet high at least and several feet in diameter. Each platter sized leaf is attached to a stem that reaches directly back down to the base of the plant. These stems appear to be sturdy but are in fact hollow and quite brittle. It's easy to overlook the young zucchinis as they are well hidden under the canopy of leaves and it is best to pick regularly when they are small. I always manage to end up with several mega squash that I didn't notice until the first frost wilted the leaves. These do well in soups and stir-fries and are delicious stuffed and roasted.

These are just a few examples of what can be grown easily and with little to no experience. Growing food is not difficult. If it was I doubt we would have survived as a species. There are some basic rules which become second nature...over time. And I do think this is the key: knowing that a perfect garden doesn't come together overnight; it takes time. The first plants I ever grew were in a garden I "inherited" in Toronto. The sunflowers were truly mammoth, so much so that squirrels would sit on the surrounding fence looking up and salivating. Ripe tomatoes fell to the ground untended, such was the profusion. My ornamental gourds were the talk of the neighborhood, far surpassing anything that had ever been seen before. When I moved to Nova Scotia I had more space. Yeah! I could plant a bigger garden. But nothing grew.

The Toronto garden had been worked and fertilized (with what, I hate to think) for years prior to me coming along and simply sticking seeds in the ground. Also, the climate in that sheltered backyard plot was far more benign than the one I moved to. But all this didn't mean a thing to me because I was completely lacking in any understanding of what plants need to thrive. I did not have a holistic view. Healthy soil, suitable climate, adequate care, pest control; I never gave a thought to any of these things so it's little wonder, after that brief flurry of success in the big city, that my first garden in Nova Scotia was a complete failure.

Certainly, one contributor to success is head knowledge, but understanding and developing a relationship at a deeper level with all the natural systems is what takes us beyond any rigid schema and integrates us as part of the process. A successful gardener is not just an outsider hopefully sticking a few seeds in the ground. Permaculture, with its clearly defined concepts of zones and guilds, cycles and interdependencies, facilitates this integration in a most soul satisfying way.

What? And When?

There will be more than one chapter on this topic because the options create a cornucopia of possibilities. Even then, much will be left unsaid because, as previously mentioned, this book is not meant to be the ultimate gardening book, a veggie planter's bible so to speak. At least one really comprehensive gardening guide needs to be a well-thumbed volume on any Permie-person's shelf. Several such books are mentioned in Suggested Reading.

A logical organization of this topic would indicate sections on early, mid and late season plantings. The concept of logic appeals to me but that other word, organization, takes me way out of my comfort zone. Not my strong suit! However, I can cite the natural overlap and interlocking complexity of the natural realm as reasons for any detours from this matrix. Permaculture is based on the cyclical rather than the linear, after all.

Several plants which need to be planted *as early as the ground can be worked,* as the seed packets often state, are greens such as spinach, chard, kale and salad greens. There are many delicious salad greens and salad mixes such as mesclun, mizuna, arugula, Bibb and romaine, just to name a few. I tend to prefer the leaf type of lettuce mixes. These are harvested by taking leaves from around the outside thus allowing the small inner leaves to mature and

✳ A typical mesclun mix thriving in the cool of late spring/early summer.

grow. This way a single plant can supply several salads. Romaine, if left, will form a head (as will Bibb, iceberg and others) but it can also be harvested as leaf lettuce from the outside in. Much as I love salad, and despite the fact that we eat it with most meals, there is a limit to how much one household can consume. With this in mind it is always a good idea to stagger plantings, that is, to plant a certain amount of each variety every two weeks. If managed well,

Bees love Bergamot (the herb that gives Earl Grey tea its distinctive taste) and I like to keep several plants blooming in my veggie plots.

Black currants thrive in our cool, damp environment. They're noted for their high vitamin C content and are delicious in pies or as jam.

These jalapeños started out green but turned red in time for Christmas. The pot hanger often comes in useful for drying herbs and peppers.

It's like watching a miracle as cracks appear and baby chicks struggle to free themselves from the confines of an egg.

These day old chicks have just been transferred to "the nursery" and will spend a few weeks under a heat lamp before being moved outside.

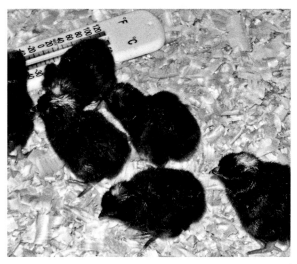

Nurse Juno finds the transfer of chicks from incubator to nursery most interesting and she insists on inspecting each newcomer as it goes by.

My first attempt at fedging might be less than perfect but this natural willow fence rooted well and does a great job of keeping marauding geese out of the vegetables.

Our newly adopted designer dog was eager to prove she could adapt well to life on the farm.

Some fascinations never lose their sparkle and perhaps this little guy will grow to cherish farming as a wonderful way of life.

As winter storms rage, the green house must be emptied and after feeding from it royally all summer there's still a glorious harvest to be enjoyed.

Look what I found in the feed shed!

Rosemary, parsley, pineapple sage, chives, Italian parsley, basil, thyme and lemon verbena all thrived just outside the kitchen door in this three tier planter.

Nigella (black cumin), carrots, tomatoes and onions are all thriving in a very small space due to the super rich soil that keeps them all supplied with an abundance of nutrients.

Not quite a food forest, this "purposeful garden" has a willow stand for harvesting, along with elderberry bushes, Haskap bushes, Jerusalem artichokes, herbs and strawberry pots, all tucked away in a semi-wild zone.

Nasturtiums gone wild! Flowers and leaves add piquancy to salads and garnish, the seeds are pickle-perfect for making mock capers and their brilliant shades of orange are simply a delight.

Sweet like candy, these cherry tomatoes needed to be rescued from the frost. Once sorted, they will store well in paper bags.

Small bags such as lunch bags work best for storing and ripening tomatoes. I need to remind myself to check them regularly.

The rabbit cages housing the breeding stock are hanging rather than free standing in the duck and goose shed, making efficient use of the space.

Creating meals that are one hundred per cent home-grown is a fun challenge, especially in the shoulder seasons, but it feels great to be able to do it and tastes even better.

We store all our feed in plastic garbage cans and they work well, except when the lids get left off.

These three drakes deserved their title of the Terrible Trio. Whatever the ruckus in the duck pen, they would surely be at the heart of it.

This wonderful sunflower planted itself and even though it crowded out a few veggies the pleasure it gave each morning was worth that. Mother N. knows best so I like to let her have her way.

Young turkeys need to be watched carefully when transferred to a new environment to ensure they can locate their water— in the large white and red container front right.

staggered planting will ensure a steady supply of salad greens well into the heat of summer. Once the weather turns hot spring greens tend to bolt, no matter how late they were planted. They will hang around longer in partially shaded areas.

Peas and beans are both legumes and are very beneficial for the soil. This is because of the bacteria which colonize small nodules on their root systems and fix the nitrogen content in the soil. This "fixing" process actually transforms the nitrogen into a form which the plants can use. Both peas and beans are available as bush and the taller "pole" or "trellis" varieties which require a sunny wall or some other form of sturdy support. Peas thrive in early, cooler temperatures whereas beans prefer the soil to have warmed up somewhat before they are planted. If you only take one hint away from this section on planting peas, let it be to install the support for the peas before the seeds go in the ground. No matter how sure

Hot weather makes cold weather plants unhappy. They want to cash in their chips and sign out, but they have a duty to perform first and, contrary to our egocentric view of the world, it's not to feed us so much as to ensure their progeny continues the tradition. In fact, it's purely coincidental to lettuce that we like salad greens. Just as surely as ducks lay eggs, a lettuce's purpose is to produce fertilized seed and once a lettuce decides to bolt there's nothing going to stop it. There is a subtle color shift in the leaves, which also turn coarser in texture and develop a bitter taste. Most notably, a tall flower spike will emerge, topped with small, unremarkable flowers. Certainly, lettuce flowers don't rate high as ornamentals and when a plant decides to bolt it might as well be pulled out... except of course for the plants designated to supply seed for the following year. Although the plant is quick to bolt, the lettuce seed is slow to form, and not always viable when saved. I always tag plants I want to keep for seed with some bright orange survey tape so I don't inadvertently pull them out.

✳ This oak leaf lettuce tried so hard to bolt. It grew a couple of feet tall but nevertheless kept giving us wonderful salads right through 'til frost. N.B. Most lettuce is not so accommodating.

you are that you will install the trellis once the pea shoots have surfaced, chances are you won't, at least not before the tendrils of several plants have intertwined beyond any hope of separating them.

It's not hard to create a support. An elegant trellis would be nice but totally irrelevant once the peas have filled in and covered it. Forked branches work as well as anything with the more twigs attached the better. Two sizes of branches are best. Short twigs that begin branching out close to the ground accommodate early growth and taller sturdier branches that start to fan out three feet or so from the ground will support the vines as they mature. They should be firmly anchored as they will be supporting quite a weight of healthy growth.

I have several four by eight foot sheets of metal grid, with square divisions of about six by six inches (reinforcing mesh, left over from the construction of the cistern). These are perfect for early growth, but in healthy, rich soil pea vines can grow up to fifteen feet high, so I usually need to install additional netting above the wire mesh to accommodate the extra growth. Last year the vines crawled along the support wires and onto the roof of the garden shed. Worth keeping in mind when the first delicate pea shoots poke tentatively out of their trench and all these directions seem so excessive.

Snow peas (named to indicate just how soon they like to be planted) and snap peas are both wonderfully tender and sweet and have edible, truly edible, pods. They are used in stir-fries and salads but don't produce a prodigious crop of full sized peas, the type used for drying and split pea soup. Later varieties are a much better choice for this purpose. Peas are quite susceptible to a withering dry mold. Once this appears it is best to pull out all the vines and

Peas do not like their roots to be warm, even though they enjoy warm sunny days once sprouted. The trick is to plant them in a shallow trench which contains a little compost or similarly enriched soil. Once the peas have sprouted, the trench can be slowly filled up, leaving the roots well protected from the heat and also well anchored and able to support windblown vines. The peas can't be planted to this depth (three to four inches) initially because the shoots are too fragile to force their way up through that much weight of soil.

burn them to prevent the spread of this fungus. As it can linger in the soil several years, it is especially important to rotate pea plantings around various locations. A tincture of Horsetail (the weed not the animal) is said to be effective as a preventative but this does require a diligent and regular spraying program that needs to be instituted at the very first signs of dry mold.

Beans like much warmer soil and will rot in the ground if the soil feels too cold. They germinate quickly and produce much sturdier shoots than peas, which is no doubt why they are often used in elementary school science projects. Other than the obvious, there are several differences between bush and runner beans. The bush type are often referred to as "snap" beans because their tender pods are crisp and will snap easily when picked early. They can be eaten raw in salads, cooked in stir-fries or served simply as a vegetable side dish. When left on the vine the pods will gradually toughen up and dry, producing beans for soup, baking or the next year's garden. There are several all-purpose beans and equally as many that are better suited for either drying or eating fresh. The "soup" beans I planted this year were disappointing as green beans but have supplied a beautiful crop of pure white beans that tumble out of their pods as though they just couldn't wait to get out.

Some runner beans, when picked young enough, will seem quite similar to bush beans but others, such as Scarlet Runners, have a more substantial pod containing larger, flatter beans. These are the beans usually cut diagonally and served or frozen as "French cut" beans. I would argue that they are more vigorous plants, provide great shade canopies and have a better flavour than other pole type beans. It is undeniable that hummingbirds love their bright red flowers, so they are certainly worth planting, if only for that one reason.

Broad beans are another kettle of legumes altogether. They like to be planted very early (some varieties in the preceding fall) and are quite slow to grow. Instead of hanging off a graceful vine the overlarge pods sprout upwards off a rugged stem and in truth, the whole plant looks somewhat ungainly. These plants definitely

✳ Broad beans tend to look like swollen fingers and the plants that bear them are equally unwieldy.

need to be planted within a substantial framework as they tend to become too heavy to support themselves. A typical garden stake won't do. The unique taste of these beans isn't for everyone but they are exceptionally nutritious and are especially satisfying in a spicy tomato sauce. They take up a fair bit of room and are probably not the best choice if space is limited.

﹡ These onions were meant to be harvested as salad greens but matured into full sized onions despite the close spacing and the invasion of chard.

Onion sets, the small yellow or white onions usually sold by the hundred in plastic net bags, can go in early, and when planted around or among other seedlings, might help to keep the slugs at bay. They will very quickly produce green onion sprouts that can be harvested without pulling up the bulb for a continuous supply. Leeks, which at a certain stage might look a bit like young onions, need to be planted very early, as they are slow to grow. They can be planted directly in the ground but I prefer to start them indoors as their delicate first shoots are easy to miss or mistake as grass.

This is where the matrix starts to fall apart. It's hard to discuss spring plantings without considering the harvest. Spring is when decisions are made, but fall is when the wisdom of these decisions is revealed, and valid decisions cannot be made without enough information, so fast forward to fall and the onion patch.

✳ Egyptian onions preparing to walk.

Egyptian onions come first simply because they're my all-time faves. These onions, also known by the less exotic but more explanatory name of "walking" onions, fit well within permaculture parameters. They multiply prodigiously and are therefore fully self-sustaining. They do also "walk" which makes them an interesting addition to any vegetable plot. As well, they require minimal attention, are frost hardy and can be harvested long after a typical onion. These versatile little onions have a taste and texture somewhere between a leek and a typical salad onion and are good in salads and stir-fries, as well as in soups.

During the first part of the growing season they look much like a healthy set of bunching onions, with vigorous green shoots that grow twelve to fifteen inches high. Then, they produce a single stem which is topped with the clump of bulbils necessary to produce next season's onions. The weight of these bulbils, which are purple/pink in colour, will gradually force the stem to bend over and gently lower the bulbils to a clear spot of earth about a step away from the mother plant. In theory, if left to their own devices, the onion plants would continue to "walk" along in this way, year after year. In actuality it's best to harvest the bulbils and plant them elsewhere in rows, shortly before the first frost.

The green tops of Egyptian onions can be harvested throughout the summer and, after the bulbils have been harvested from the top of the stem, the smallish onion which forms the base of the mother plant can also be dug and used. These onions and their

✳ Egyptian onions cleaned and ready for use.

green tops are very cold resistant and, with a protective covering of straw, can be harvested as long as the ground can be dug. In the kitchen they require a bit of preparation as each onion comes well wrapped in several layers of protective skin. However, it's worth the bit of extra effort to have fresh greens dug in December in a place where the first frost can come as early as September.

Leeks are also a member of the onion family. They winter fairly well in the ground but are totally dissimilar to Egyptian onions in their planting requirements. They are a very slow maturing crop and it really is best to start them indoors. Leek seeds are quite small and even with an efficient "mechanical" seeder it is difficult not to over-plant (read overcrowd) the delicate, hair-like baby plants. Thinning them is crazy-making! Also, because the young seedlings look just like blades of grass, young leek plants are quite likely to be weeded out of the garden by mistake. It's much easier, and definitely more efficient, to sit inside on a cold February day, relieving that itchy green thumb by pressing individual seeds into cell-packs, and to wait for a couple of months before introducing them to the garden. Not quite as simple as planting potatoes but the first taste of leek and potato soup will prove the effort worthwhile.

While in the fall garden, we might as well look at a couple of other vegetables that can be dug late in the year but require spring planting: beets and parsnips. Parsnips especially are better after they have been touched by frost. They are not difficult to grow but for some reason are not as popular as they might be. The white carrot-shaped root can be sliced lengthwise, breaded and fried or they can be boiled, then mashed with carrots or potatoes to give a "sweeter" more complex flavour. They can also be added to stir-fries or stews, and curried parsnip-apple soup is truly delicious. Parsnips need to be seeded directly into the ground early in the spring. They are very slow to germinate so it is important to mark the place quite clearly, and remember to be patient.

Beets are not quite as resilient. They will endure a certain amount of light frost but not a heavy freeze. They are one of my staple crops as they are incredibly nutritious, versatile and easy to grow. Early beet greens with a dab of butter. Yum! And borscht soup is a classic for good reason. However my favourite way to serve the fresh dug beets is as a warm topping for spinach salad; glazed with a honey balsamic dressing and mixed with lightly candied walnuts, all topped with goat cheese. Enough to convert any reluctant gardener into a fanatic!

Beets germinate fairly quickly—in 10–15 days. The seed is about the size of a shrivelled up pea and therefore relatively easy to plant, usually in carefully spaced squares or rows. However, because fresh beet greens are so very tasty, it's not a bad thing to seed generously, planning to include a few servings of beet green "thinnings" on the early summer menu. What's neat about beets is that they can be left in the ground well into the fall and will store quite well in a cool damp place for several more weeks. They make excellent pickles, are a good source of iron and noted as a diuretic. Definitely worth planting!

✳ How to get the beets in the jar? Hmmmm! Logistics never were my strong suit.

✳ Chard and beets are still happy in late November.

All the veggies mentioned so far need the mild days and cool nights of spring to germinate and become established. As already mentioned, many of the crops that thrive in the comparative coolness of spring or fall will "bolt," that is, go to seed in the heat of summer. Lettuce and spinach provide two perfect examples of this. Spinach, the more hardy of the two, can be started very early, as soon as the first hints of spring have the birds all atwitter. If the earth has been put to bed properly in the fall, well-dressed with layers of compost and manure (and/or seaweed), it's just a matter of removing some of the straw mulch, scratching the surface of the soil and gently pressing in some seed. It really is that simple. End of March, early April is not too early to try this. If, worst case scenario, there is a prolonged cold spell, the seedlings can be protected by pulling the mulch back around them.

Many vegetables, and especially root crops, are best seeded directly to ground. Sure, there are methods of wrapping seed in wet paper towel to speed germination, but introducing fragile seed sprouts into the ground is tricky and time consuming. It has also been my experience that direct plantings, which have not been shocked in any such way, usually seem to catch up and surpass the seeds that have been "tricked" into growing. Remember that one of the rules of permaculture is to strive for maximum gain from minimum expenditure (of effort in this case). I prefer to keep things simple and plant directly to ground whenever possible, keeping in mind that the short cool growing season in these parts rarely allows enough time for slow maturing plants to be seeded directly outdoors. Such plants require an indoor start, whether at the local garden center or in our own greenhouse. Did someone just say Greenhouse?!

7

Gotta Getta Greenhouse

I used to view a greenhouse as a luxury of the first degree. Now I realize they are actually an integral part of a successful garden. In essence, a greenhouse must allow in as much natural light as possible, while protecting plants from extremes of temperature and precipitation. In the spring and late fall it's necessary to keep plants warm but in summer it becomes equally important to keep temperatures from rising too high. A thermometer that records both highs and lows is essential for doing this. Such a thermometer has a round dial face and three hands, very much like a typical clock. One hand registers the lowest temperature it has dropped to, one registers the highest temperature it has risen to, and the third hand marks the actual temperature in the greenhouse at any given moment. They are not terribly expensive and are so useful in monitoring the temperature fluctuations within the greenhouse that I would class them as indispensable.

✳ A high/low thermometer is indispensable in a greenhouse.

A greenhouse has a special feeling all its own, whether on a cool spring day with rain pattering on the roof or a late summer day when tomato vines are threatening to poke their way through the roof. One of my earliest childhood memories is being shown my uncle's greenhouse. I was not allowed to enter but the spicy, exotic smell of tomato vines was enough to hold me wide-eyed

❋ The exotic smell of tomatoes growing in a greenhouse lingers in the olfactory memory, perhaps even for a lifetime.

and enthralled as I peeked around the rickety old door. Even to this day that smell transports me back through time and across the ocean to the grimy industrial north of England and my uncle's greenhouse. Perhaps that's where some of the seeds for my own life's journey were planted. Who knows?

Our greenhouse is a simple frame affair covered with a heavy duty, semi-transparent plastic. It measures sixteen feet by nine feet and didn't cost much over a hundred dollars to construct. I was fortunate enough to be in a large hardware store when several rolls of translucent plastic tarp were being dragged out from some dusty corner of the storeroom. As they were not "in their system" no one knew what price to put on these rolls of poly sheeting, so they were priced ridiculously low for a quick sale. For once I was in the right place at the right time and I bought all that was available, knowing it would come in handy for any number of projects for years to come. Without this lucky purchase we probably would have used

✴ This greenhouse was simple and inexpensive to construct and has survived several years of intense weather patterns.

a heavy gauge vapor barrier, which I'm sure would serve the purpose equally well, or special purpose greenhouse film which is also available—at a price.

There are many really good designs for greenhouses available through university extension offices and it's best to see what's available before deciding which is the best fit for the space and requirements it is expected to fill. I would encourage starting small and moving slowly, not just with greenhouses but with all aspects of development. This will help to keep the learning curve from feeling like a roller-coaster ride!

We recently attended a workshop on "hoop houses," which are large, commercial (market-garden sized) greenhouses and which are very economical to purchase and construct. They have steel ribs which support a "skin" of specially treated plastic that has a UV protective coating on the inside, giving it the magical ability to bounce sunlight back onto the plants for prolonged benefit. Certainly worth checking out for anyone who wants to jump in with both feet but really, it's better to start small so that mistakes and crop failures are similarly sized. From this perspective, our smaller, less advanced, basic greenhouse presents itself as an infinitely more user-friendly space. Here on the coast we get phenomenal wind pressure at times but we've only had to repair the plastic cover once. I believe the key to success here is to have the plastic stretched as tight as possible and to reinforce all contact points.

Probably a greenhouse constructed with recycled windows would be more in keeping with permaculture ideals and perhaps it would be more appealing to look at, if some thought were given to its design. Traditionally the main requirement for any greenhouse was that the structure ran east to west, giving maximum southern exposure, and that light should be able to enter from the roof and from the south and east and west facing walls. There is some debate as to whether this can cause overshadowing of plants on the north side, especially in winter when the angle of the sun is low. A row of taller plants or vines along the southern interior wall will to a certain degree shade plants to the north of them, even in a greenhouse with a clear roof. Also, the shade from exterior sun-blocks, such as trees, will change seasonally. I only mention this to suggest that the traditional east-west axis of a greenhouse might need to be adjusted to accord with the main purpose, which is to allow for as much natural light as possible. This is where detailed zone and sector maps (see Chapter 9) are so useful. A greenhouse can cover workable ground with plants growing directly in the soil or it can house pots and seed trays, raised beds or a combination of all three. Presently all our greenhouse plants are grown in pots or seed trays.

Our choice of where to build the greenhouse was limited by the availability of flat ground. Planting directly in the ground was not an option because of poor soil quality in this area. In time we will probably construct a raised bed down one side but presently we rely on large pots. This allows for a certain flexibility of movement in the early growth stages but the downside is that pots dry out quickly in the heat.

Various irrigation methods are available for greenhouses. I was fortunate enough to notice some sizeable fiberglass troughs in the "bone yard" of a local business depot. Overgrown with weeds, they'd obviously been sitting there for a while and when I asked if they might be for sale at a reasonable price the proprietor was delighted. He had a massive cleanup underway and was wondering what on earth he was going to do with this unclaimed custom order. It never hurts to ask! These troughs make perfect baths for the growing pots to sit in and, while making watering so much easier and efficient, they also help to guard against things drying out. On the downside, standing water can contribute to mildew and rot (a major problem in any enclosed growing space), not to mention mosquitos, so while very useful the troughs need to be monitored carefully. It's always good to be on the lookout for the less obvious alternatives and these troughs are a good example of that. Remember, in permaculture there are no problems, only creative solutions.

✳ In the early spring the greenhouse provides protection for tender seedlings which will be moved outside as the weather warms up. The recycled troughs keep the seed trays moist.

✳ These heat loving plants; peppers, tomatoes, aubergines and cucumbers, will thrive in the greenhouse throughout the summer.

Once the greenhouse is established, what to put in it? Broadly speaking there are three main purposes for a greenhouse: for starting seeds; for growing plants that require a warmer, more protected environment than outside; and for prolonging seasonal growth. Some seeds are just not meant to go in a greenhouse as they are hardy, not fond of too much heat and prefer to go directly in the ground—beets, carrots, parsnips, spinach or chard for instance. On the other hand I wouldn't consider seeding cabbage, Brussels sprouts or leeks directly to ground. They need the extra time that pre-seeding indoors gives them. Other plants such as tomatoes, peppers, aubergines and cucumbers will complete their whole cycle in the greenhouse because in coastal Nova Scotia temperatures fluctuate too much, even in mid-summer, to ensure ripening.

One difficulty is to decide exactly when to start pre-seeding. I usually leave it perhaps later than I should because in the past I've ended up with trays of gangly seedlings, trapped in too small seed cells, that can't be put outside because there's still snow on the ground. Certainly not before the end of February, I've promised myself, while secretly deciding that mid-March is a safer bet. Another thing I've promised myself is not to overplant. This is easy to do when a scant tablespoon of seeds might plant more than five or six seed flats. How much cabbage are you likely to eat? Of course, by the same token, how many plants are the slugs going to eat? It never hurts to have a few extras on hand and if all the plants prosper they can easily be gifted or sold.

Brussels sprouts are members of the Brassica family, along with cabbages and broccoli. They are all best started indoors because they're fairly slow to mature. Once established, they are very hardy but unfortunately cabbage white butterflies find them irresistible. They lay eggs on the plants which hatch into little green caterpillars. These feed on the plants before continuing their cycle, becoming cabbage whites and laying their eggs.... Without a doubt the best way to avoid this problem is to cover the transplants as soon as they are set out with some kind of filmy row cover which

* Brussels sprouts are best started inside because of their slow growth rate, but definitely need to be moved outside as soon as risk of frost is over, as they don't like excess heat.

will allow light and air to penetrate while deterring any flying predators. Commercial row cover is available but discarded window sheers will work equally well.

Brussels sprouts are quite fun to grow as the miniature cabbages form where the stems of large unwieldy leaves attach to the plant's tough woody stem. In the fall the leaves wither and begin to fall off, leaving a pyramid of Brussels sprouts. They are very hardy and the flavour is improved by exposure to frost. It surprises me that more people don't grow Brussels sprouts but when I voice this opinion it's usually met with rolling eyes and condescending sighs.

What to plant in also requires some thought. I remember seeing a nifty little gizmo in a tool catalogue that was designed to make seed cells out of newspaper. This is probably the greenest option but I must admit I've never done it. We don't get the newspaper is my default reasoning, but secretly I'm not sure that I'd have the patience, or the time, to sit and fold newspaper strips around a wooden cone night after night.

We tend to favour compressed peat pellets that look like miniature hockey pucks until they're introduced to water, which causes them to swell into adequately sized seed cells. The nice thing about these is that they're made of compressed sphagnum moss. This allows the tiny roots to spread with ease, while absorbing sufficient water. They can be planted directly into the ground without disturbing the sensitive root growth which is a great plus and should be kept in mind when considering the options: plastic but re-useable versus organic and way more plant/planet friendly. Whatever your choice it's always good to shop around as the prices vary considerably from source to source. Suppliers know that a dash of spring fever plus a severe case of itchy green thumb can cause any parsimonious budgetary concerns to float away on the first mild breeze.

Once the seeds are well sprouted they might need to be put (still in the seed pellet) into bigger (three/four inch) pots if a huge amount of root growth is sticking out the bottom and the soil outside is still too cool to plant in. Such a scenario is unlikely to happen with leeks, for instance, because of their slow germination time but it could easily happen with squash. Leeks therefore would be seeded several weeks before squash. There's always some trial and error but a good gardening book or two will help to eliminate some of it. Most importantly it's best not to expect one hundred per cent success from every single seed that's planted. Mistakes will of course be made and conditions will vary, so what might not grow well one year may thrive the next.

Two things that might not initially come to mind along with those first bright greenhouse plans, are the need for adequate ventilation and airflow, and access for insects, both beneficial and not so. If there is not enough airflow the plants will become prone to mildew, and once any sign of mildew is noted the plants must be removed and destroyed. It's a common problem and can become virulent. We have a door at the east and west ends of the greenhouse and there is a vent above each door that stays open even when the door is closed. In damp or overcast weather this is not enough, and we need to install a fan and rethink the vents. We definitely underestimated the need for consistent air flow, especially in the extra moist climate that prevails here.

Insect access is another need that mustn't be underestimated because if flowers aren't pollinated they won't produce any fruit. The scarcity of bees has been noticeable for a couple of years now, which touches on the topic of installing a beehive. More on that another time, but not in this book. On several occasions Calum has spent considerable time dressed in a fuzzy yellow and black striped bee costume, fertilizing the greenhouse plants with a fine sable brush.* This method does work if you have more patience than I do.

Of course, open access to insects allows for less welcome visitors. Aphids have been our main problem for the past couple of years. They go crazy for aubergine plants. When researching what to do about aphid infestations I came across one site that recommended installing a couple of aubergine plants to distract the little pests from the rest of the plants. Hmmm! While I'm sure this is a perfect solution for anyone who doesn't particularly care about

---

* I doubt this will make it through the first cut editorial comments and I'll probably have to admit that he was actually wearing jeans and a tee-shirt. The bee suit? Well, if I was a flower...

✳ These young aubergine plants will attract aphids like crazy. Great if they're intended as "sacrificial" plants, but not so much if the intended purpose was a harvest.

aubergines, it really wasn't much help to us. Ladybugs work better. Aphids to ladybugs are like chocolate to a chocoholic. Good garden suppliers actually sell ladybugs by the hundred. Of course, there's no guarantee the ladybugs will do the honourable thing and stick around once you've purchased their freedom. In truth, I expect they pigged out so much that they couldn't fly off even if they'd wanted to, but we'll just say they were being loyal.

Maintaining the equilibrium of a greenhouse truly brings into focus the importance of natural balance. Certainly, the aphid-ladybug connection underlines the value of keeping part of any property in its natural state, to allow the proliferation of natural predator-prey relationships. No matter how small the area being worked, any successful permaculture plan requires a natural zone, usually labelled as zone five, the furthest perimeter of the bullseye which has as its center the heart (or house) zone, called zone one. Zone one can be viewed on an inspirational level as the intent to live in harmony with nature, rather than to attempt to dominate it. Zones two and three are also planned around human activity: which garden beds will be visited the most regularly, what's the most convenient placement of the chicken house, the compost bins? These are the considerations for areas closest to home whereas zones four (food forest) and five (wild zone) are where the vision or intent of zone one is honoured in the naturally occurring systems left to flourish untouched.

Warm Wussies

Around these parts May 24th was the traditional planting day. Being a long weekend it was convenient and in theory the weather and, more importantly, the soil had warmed up by then. Not so anymore. It seems almost as if the seasons have shifted, installing June as the "new" May and October as the "new" September. Before I knew any better I used to plant everything before the end of May. No wonder my successes were usually outnumbered by my no-shows and wimp-outs!

Seeds that demand warm soil will rot in the ground when soil is too cool. Transplants that are put out when the soil is too cool will be so shocked that they are liable to freeze (metaphorically) and maintain a position of complete stasis, neither flourishing nor dying. Sometimes it's hard to resist the urge to plant on a warm sunny day in May but if the ground has not had a chance to warm up sufficiently and the nights are still cool, precious growing time will be lost rather than gained and the need to replant will be a waste both of physical and financial expenditure.

Taken out of context the foregoing paragraph would be a complete contradiction to statements in the previous chapters regarding seeds that require cooler temperatures in order to germinate. It's all dependent on the personality of the plant. It might sound

✳ One garden bed—three "time zones." Egyptian onions (center) have already produced their seed, whereas the Zucchini (right of center) is just starting to fruit and the Brussels sprouts plant (lower right) will grow another foot or so before it starts to mature.

complicated to a beginner but it really isn't, because knowing who likes what quickly becomes second nature. Broadly speaking, beans and squash, including cucumbers, must have warm soil. Potatoes are not quite as tender but a cold spell in late April or early May can set them to sulking (refusing to sprout) or worst case, turn them into a slimy mess.

Two things worth mentioning here: phenology and solarizing. Phenology is a method of gardening that goes by nature's calendar rather than the Gregorian system of numbered days and months, which really has no connection to weather patterns and temperatures. Phenology uses moon phases and natural growth cycles to dictate garden rhythms. For instance, Forsythia blooming will indicate it is time to do some early planting. With this system the gardener "reads" instructions from the surrounding flora which are finely tuned to that specific locale. I love the concept but only know enough about this system to get me into trouble, so I'll say no more. This topic requires a book unto itself.

Solarizing is a way of heating up the ground artificially by covering it with black plastic sheets. It can also be used very effectively for killing weeds (and everything else) in a given area. The heat that is gathered on the black shiny surface, when transferred to the ground, literally burns the weeds up.

A great source of rugged black plastic sheets is the local lumber yard. Certain products come shrink wrapped in this plastic which is usually at least twelve feet long and almost as wide. It usually gets bundled into the garbage hopper and so, from a permaculture standpoint, taking advantage of this free resource gets pretty good marks, for reuse of product. It also eliminates the need for those Agent Orange style weapons still being marketed to eliminate the dandelion nation. I doubt that anyone interested in reading this book would be remotely interested in using such products. However, one easy way to quiet the inner-child who stills longs to be a super-hero is to become Super Solarizer, swooping in to aid friends and family alike who are troubled by *those unsightly weeds in their walkway.* Another way is to take them a bottle of dandelion wine.

Constructing garden beds using the hugelkultur method, as mentioned in chapter three, is another way of ensuring warm ground. The heat generated by the microbial activity in the rotting wood fibers can make a difference of several weeks, to the advancement of beans, squash and potatoes. This is going to vary according to the stage of decomposition the hugel is at. I've only been working hugels for a couple of years so this is not the voice of huge hugel experience but when I have done early trial plantings, thinking I've been really pushing my luck, they have been successful.

One way to be sure the ground is warm enough, without wasting too much seed, is to plant just a few seeds, leave them for a week or so, and then dig a couple out to see whether they have sprouted or rotted. Of course by this time a week or so has passed so the earliness factor is becoming somewhat irrelevant. More often than not plants tend to catch up when conditions are right and come harvest time it's usually impossible to tell the early plantings from the later ones. With this in mind I prefer to err on the side of caution and plant when I'm sure the soil is warm enough.

The main reason to strive for as early a planting as possible, other than to satisfy the longing for fresh-out-of-the-garden veggies that has been swelling all winter, is to maximize the growing season. No matter how healthy a plant and no matter how perfect the weather there's a certain time requirement. Each plant has its specific number of days to germination and days to harvest. Some gardeners can gauge this growing time very accurately using the "degree days" calculation method which takes into account the hours of sunlight and the angle of the sun, among other things. This method works well with phenology and one day perhaps I will take the time to study them both more fully.

Some plants such as Brussels sprouts need well over a hundred days to harvest time whereas other crops such as radishes require less than sixty. These numbers are influenced by time of planting and type of weather. Even the degree of a south facing slope has a significant influence, with approximately ten degrees of incline being optimum and contributing to a significantly earlier harvest.

Whether a plant is frost tolerant or not, combined with the number of days it requires to reach maturity, are the main factors determining whether or not the plant should be grown in the greenhouse or in the garden. For example, tomatoes rarely have time enough to ripen on this little island and there is a limit to how much green tomato chow, mincemeat or salsa any one household can consume, especially when they are not high on the faves chart.

For us it's a no-brainer, tomatoes are grown in the greenhouse, where the extra heat promotes accelerated growth. We also grow our cucumbers and eggplants in the greenhouse. Cucumbers and tomatoes grow and ripen perfectly well outside, only a mile or so inland, because of the extra couple of degrees of warmth. Subtle nuances in climate are just as important to consider as the broader climate zones. Also, fluctuating global patterns, not to mention warming, can easily negate the previous statement. In fact this happened last year when the yield from tomatoes and cucumbers planted outside surpassed greenhouse yields. Every year is different and that's so important to keep in mind. What flourishes one year might disappoint the next, and vice versa.

✳ Early green beans flourishing in a sheltered, south-facing cold frame.

✳ This cucumber is very happy in the warmth of the greenhouse.

Even though an area might be classed as climate zone four or five, it is possible to establish microclimates (apart from the obvious greenhouses and cold frames) which can be dedicated to warmth loving plants. For example, the area in front of a south facing wall will be much warmer than similar space by an east or west facing wall. If the south facing wall is also protected from any prevailing winds and is constructed of heat absorbing brick the effect will be magnified several times over.

Naturally occurring microclimates can be easily enhanced with additional protection from wind or perhaps the elimination of some light blocking branches. This is why in permaculture it is so important to walk, sit and observe the land you intend to work, and really experience it at an intimate level. As well as a typical map type plan, permaculture encourages the construction of a chart, somewhere between a pie chart and a bull's-eye, that describes influencing factors such as prevailing wind and path of sunlight during the year. This is called a sector map, and is a definite aid to the planning process. Truly, a little pre-planning can save a lot of re-dos. (Wow! Did I say that?)

The positioning of our greenhouse was dictated by the fact that there was very little ground left that was flat enough after the hurricane tore through uprooting trees. We built it in about the only place we could at the time, which was south of where the berry patch is now. Fortunately, it created a really neat little microclimate by blocking out the onshore breeze which picks up just about every day in the summer. Normally, a structure will create

✳ The berry patch is protected from prevailing on-shore winds by the greenhouse.

a shade zone to the north of itself by blocking out the sun, but because the greenhouse is fully translucent the light is gently filtered but not blocked out altogether.

Trees make very effective wind breaks and as we develop more garden beds, at the moment in the form of hugels, I am careful to leave patches of baby spruce in areas that are strategic for blocking the prevailing winds, which in our case blow from both north and south. Obviously, these naturally occurring but non-transparent windbreaks will in time block sunlight if they are allowed to grow unrestricted. It's important to make sure that they are stepped back from the garden so as not to directly overshadow or impede access, and also that they are kept trimmed to a manageable height. If these natural windbreaks are interplanted with wild berry bushes, such as rose and bramble, they move closer to permaculture ideals by fulfilling more than one purpose. As well as creating a windbreak they will supply natural habitat and hopefully attract more beneficial birds and insects to eliminate garden pests.

This does bear repeating as one of the hard and fast rules of permaculture. Within any given division of property a certain percentage of the space must be given over to natural habitat to allow for something close to perfect equilibrium to be maintained. This natural habitat will encourage the proliferation of beneficial birds and insects. In theory, and if permaculture planning was all done on paper and not influenced by the existing natural environment, this "wild" zone would be furthest from the dwelling along the outer perimeter of a perfect bull's-eye. Working in towards a central point, zone four would accommodate areas requiring less visitation; the food forest perhaps. Zones three and two would work incrementally toward areas of more regular use, for example, the herb and salad gardens which might well be visited once or twice a day in season and so would be conveniently placed in zone one. (I like to keep several herb pots right outside the kitchen door.) However, potato and squash beds, which once planted and mulched need very little attention before harvest time, are better placed in zone three.

✳ These herb pots by the kitchen door are so convenient and get well used.

Of course the "bull's-eye" visualization which is often used to explain zones very rarely applies to the real world. The positioning of our natural zones has been influenced by scant patches of healthy tree growth that we wanted to preserve and the positioning of wind breaks. One area that had been designated as part of the duck pen turned out to be rife with partially buried shards of shattered windows. It soon became obvious that it couldn't be cleaned up so we fenced it and allowed it to grow wild. It is now full of wild blueberries, flowers, mosses and an Indian Pear tree that is covered with blossoms every spring. Another purpose of wild zones I have come to realize is to remind us of the diversity and exquisite beauty that will flourish when allowed.

Quack, Cluck and Gobble

Now to the noisy part! As already mentioned, we keep chickens and ducks. We also have geese, turkeys and meat rabbits. Of course it is possible to design a permaculture system without livestock; permaculture principles can be used to organize a kitchen, an office or a minuscule urban lot, all places where livestock might not fit in so well. (Here I could tell the story of a sheep called Wayne who got loose in a museum, but that would be way off track!) However, in our situation, on almost an acre with extensive gardens, I can't imagine not having a totally reliable source of manure—and delicious protein to go with all the veggies.

In the early days of our journey along the Permie Path, finding a source of good manure was often difficult. For the most part, farmers want to keep the manure their animals produce and, generally speaking, horse owners seem to be the only ones who don't know what to do with the manure their horses produce. Before I became an "expert" on poop—that is, before I read *Holy Shit* by Gene Logsdon—I guess I thought all poop was much the same. Not so!

For a couple of reasons horse manure is probably the least desirable, in my opinion. For example, there is the issue of medications and external treatments used to combat the various conditions horses can be prone to. My only experience is with race horses, but it's certainly enough for me to never go near manure coming from a race track. Hobby farms and equestrian centers are a much more likely source. A good horse owner will keep the barn clean by regularly changing the bedding, which results in more straw (or more usually, woodchips) than poop. Also, because horses do not digest their food as well as ruminants, many of the seeds they consume come out the other end, well primed and ready to sprout. Horse manure has the ability to introduce a myriad of hitherto unseen weeds into a garden. In theory, this would not happen if it was composted thoroughly, with enough heat to neutralize every single seed. Colour me skeptical on this eventuality.

I know, I already said this; how many times? But…most manure needs to be composted, or at least left to sit and "cool" down for a year, minimum. Chicken manure is particularly "hot." Yes, it literally "heats up" during decomposition but it's a chemical process which "burns" tender plants. Although chicken manure can be destructive, when properly aged the results it produces are awe inspiring. Years ago, in our pre-poultry days, I had the opportunity of shovelling out a chicken shed; hard labour in return for (the end result of) chicken feed. It was a disgusting, gut heaving experience and I swore I'd never, ever, ever have chickens. Hah!

What I discovered when we got our first batch of hens was that a properly designed and maintained chicken coop is not full of squishy stenchables, but is in fact carpeted with a hard packed, virtually odourless, straw mix. When not outside pecking around, chickens like to roost, that is, perch on a horizontal cross beam suitably sized to be easily gripped by their feet. When evening comes and they're bedding down for the night there's a certain amount of jostling and muttering that goes on before everyone agrees on the lineup and settles down. Here's the important point: most of the manure drops through a wire grating under the roost into a separate box that can be shovelled out quickly and easily, once or twice a year. It's also important to decide approximately how many chickens you plan to keep (which requires knowing how many eggs you want to eat, sell and give away) and then design their coop accordingly.

The other common deterrent to keeping chickens is the Rooster from Hell. It's surprising how many people were traumatized in their youth by this particular character. I had assumed that the passing of time and the magnification of childhood memories must have exaggerated the actual menace of these legendary killers, until we inadvertently raised our own feathered fiend. After living (but only for a short time) with this guy I could clearly understand how cock fighting came into being. A vicious rooster is truly frightening and dangerous. There is only one cure for a vicious rooster: an axe.

✳ What makes a rooster turn mean? Possibly, having to continually fight for dominance.

We now have a rooster who eats out of my hand and enjoys all the perks of hero status, both from his humans and also from his harem. He managed to fight off a racoon one warm summer night, saving all his ladies from a horrible fate. By the time the ruckus woke us up, Rudi (the rooster) was a bloodied bundle of feathers, crumpled in the corner of the run. We didn't hold out much hope for his survival, but he's the quintessential tough old bird. After a few days in ICU, under the kitchen table, his progress slowed and it seemed that he wasn't going to make it after all. That's when we decided it was time to let those mother hens take over. It was funny and also very touching to see them all clucking around him as he quickly staggered back into his cock-of-the-walk persona. Much of his comb and upper beak never did grow back after The Fight but his hubris, not to mention his libido, is alive and well, and he is loved by all.

What makes a rooster turn mean? I believe, and this is just my theory, that it's all about dominance. We raised our hell-bird along with four other roosters and the pecking order was established fairly early on. There was no doubt who was bottom of the heap and second to bottom took solace in keeping this status quo. The fight for top, between one alpha wannabe and the other, became increasingly frenzied and even when it seemed to be settled there was always one usurper or another trying to stage a successful coup in the coop. I believe this is what developed the evil side of our Rooster from Hell. When we raised Rudi we eliminated any competition early on, so he was able to excel at being a lover, never needing to be a fighter.

Of course, it's not necessary to have a rooster unless you intend to proliferate your own flock. There are good reasons to do this but for me, most importantly, it eliminates the risk of importing disease from an outside source. Also, chicks can be a source of income, especially if they are a heritage breed.

In urban areas it might be better not to have a rooster because their quintessential call to the rising sun can provide cannon fodder for any local chicken wars. Some cities allow chickens, some don't. It's always good to check your zoning. Deciding what breed of chickens you want to raise is another issue which requires careful thought. There are many different "styles" to choose from. It's almost as hard as buying shoes! Going to a poultry show is one way to start but beware, there will be many magnificent looking birds fluffing their feathers at you and sometimes looks can be deceiving.

It's probably best to first decide what you want from your chickens. Is it eggs, meat or the best of both? This is an area of compromise as there is no perfect bird. We wanted to raise heritage breeds, preferably endangered breeds that were primarily for egg production. We eventually settled on Barred Plymouth Rocks (a dual purpose breed) and Brown Leghorns. If you're not planning to hatch and raise purebred chicks you could easily try a broader selection, but if you do want to raise your own chicks, a word of warning: it's very easy to be tempted to raise several

✳ We raise these two heritage breeds primarily for egg production—
Brown Leghorns and Barred Plymouth Rocks (the latter being more
of an all-purpose breed).

different breeds at once. After all, what's a couple of extra incubators, right? Keep in mind that each breed needs to be isolated for at least three weeks, with their own specific rooster, before hatchlings can be guaranteed pure-bred. This might explain why some chicken houses have the amoeba-like habit of growing additional cells on all sides. Having witnessed some nightmarish situations like this, I'm inclined to side with the less-is-more principle of permaculture on this particular topic. Also, it's good to remember that all those cute little chicks rapidly grow into teenaged marauders. I've had my own nightmare situations relating to a basement full of feathered young'uns, all with an unnatural propensity to escape their carefully constructed nursery pens. Not good! It really is best to start small and move slowly.

In praise of chickens, I will say this: even on a cold, stormy day (and believe me we do have our share of storms here along the coast) after gearing up and fighting a path out to the animal sheds, there is something innately satisfying about chickens, tucked safely in their straw, bumbling a muted "thank you" for the feed and hopefully providing a few eggs. It's difficult to explain this logically but I believe that by reaching under a chicken for a warm egg

✳ Morning sun on fresh laid eggs is the best start to any breakfast.

we also reach deeper within ourselves, satisfying a primal need to be more intimately connected with our food, honoring its provenance, rather than simply taking it as something sterile, from a cooler, in an over-lit food mart.

This need for closer connection and deeper understanding goes beyond food and in a way represents the heart of permaculture, as the need to integrate with natural systems rather than to remain distanced from them, as modern society would have us do.

Chickens (and this could be extended to poultry in general) are often used as examples of a perfect closed cycle, from a permaculture point of view. They consume kitchen scraps and garden waste and supply eggs and meat. They provide manure which in turn will be used to create more kitchen scraps and garden waste. The eggs they produce feed us but also create more chickens. There is no waste and the process is cyclical and ongoing.

We expanded our poultry operations to include ducks when we discovered that they were the ultimate slug control. Once again, choice of breed depends on whether the main purpose is for eggs or meat. We went for eggs and chose Khaki Campbells and Indian Runners, both noted for their prodigious egg laying capacity.

Recently we had a momma duck who was a cut above the average and soon noticed that the clutch of eggs she was working so hard to produce never swelled beyond one or two (due to the daily visits of those pesky people with the egg gathering basket). She took the initiative of leaving the safety of the duck shed to build her nest out in the big wild world. When she disappeared we thought she must have been taken by a fox, until we finally noticed her new nest, tucked away under some berry bushes in the front yard. It was so well hidden that even though we had been walking by it several times a day it had remained unseen for a couple of weeks. Momma duck sat motionless, snuggled into the mulch of dry leaves, never leaving her eggs until they hatched; all nine of them. Incubation time for duck eggs is twenty-eight days plus the time it takes to assemble the clutch. That's dedication! What was even more heartwarming was the fact that she knew her babies

❋ Chickens personify the closed cycle, multi-tasking principle. They
lay a plentitude of eggs for food and resale and produce fertilizer
to grow vegetables; offspring are sold for income and meat is
consumed. The flock reproduces itself, they recycle kitchen waste
to create compost, and de-bug and aerate the vegetable beds to aid
better vegetable growth, leftovers from which they will eat.

wouldn't be safe where they were. When they were a day old she
marched them all back to the comparative safety of the duck shed.
Far be it from me to contradict The Bard, but in truth it's a momma
duck defending her young and not a spurned woman that hath

* Momma duck protecting her young.

fury exceeding anything found in hell. Her curious flock-mates very quickly learned to stay several feet away, or else!

Ducks are a wonderful addition to our homestead even if they are a little harder to keep because of their water requirements. Unlike chickens and turkeys, who are quite happy to have their water dispensed in a shallow trough that can be fitted with a heating element to prevent freezing in the winter, ducks require at least four inches of water so they can submerge their heads, not just their beaks, when they drink. This keeps their nostrils and eyes clear of mud and silt gathered while foraging for slugs, etc. They also like to play around a bit and have a bath after they've finished drinking. (Imagine the bathroom after a couple of three year olds have had playtime in the tub.) In the summer months, no problem, but in the winter when everything freezes solid in an hour or so, I often wish they weren't so messy.

So why keep ducks? Well, definitely to decimate the slug population, but we also prefer the taste of their eggs which is slightly more buttery than chicken eggs. Athletes in training seek them out

✳ Ducklings surely are the most adorable of hatchlings.

because of the additional nutrient content and bakers claim that they produce superior results. As if this weren't enough incentive, of all hatchlings, ducklings are without a doubt the most adorable.

A very close second in the cuteness category are the baby Californian White rabbits which we raise for meat and also for sale as breeding stock. It amazes me that more people don't produce at least some of their own meat by raising rabbits. I was totally unfamiliar with rabbits until we started raising them, but somehow had the notion that they were scrawny, tough and strong tasting. I was mistaken on every count. There's a lot of meat on one rabbit, enough for at least six or seven generous servings, and it cooks and tastes very much like chicken. The great thing about rabbits is that they also provide lots of fertilizer that can be put directly onto the garden. It's the one manure that should be used directly and should not be composted, as this degrades its quality.

The turkeys we raise, Beltsville Small Whites, we chose for meat (obviously) and also because they are a heritage breed that is endangered. As it happens they are also quite prodigious egg producers, but only for a couple of months during mating season. They're very easy to keep but you really do have to think for them. I'll say no more.

For years I was a confirmed vegetarian so it seems strange to hear myself enthuse about the quality of meat. When we started raising animals for meat I realized it was in some way hypocritical of me to produce but not consume. I felt I needed to be part of the cycle I was creating. Also, I was beginning to question the wisdom of a diet that contained no animal protein. I still don't eat any meat that may be in any way connected to factory farming, in other words anything purchased from a food mart, but I do occasionally consume animal protein we have produced ourselves because I know that animal was honoured and had a good life.

✳ A Beltsville White tom in full mating regalia. He complements this look with a deep throbbing vibration that sounds like a distant sub-woofer.

Now, I come to the geese. In truth I have no idea why we have them, other than to preserve yet another endangered heritage breed, Pilgrim geese. They are docile, compared to other breeds, and they are the only breed of geese whose gender can be differentiated by their colour. The males are white with blue eyes. The females are predominantly a brown/grey mix, with some lighter flecks. I have read accounts of geese being used to weed strawberry patches but our geese certainly didn't excel at this task! Geese do have the endearing quality of mating for life and to see them strutting their stuff in the spring is cute, but I don't think I could put up a convincing argument in defence of keeping geese.

(The last statement sparked a vigorous response from my husband, and a convincing argument, as follows: they taste good, produce big eggs of the best kind and are raised on only pasture, with no additional feed needed for most of the year!)

While geese tend to strut around imperiously, casting a haughty glance at anything that might happen to be in their path, chickens scratch and dig diligently. There's nothing they like better than to be let loose in a garden. I tried letting them out in my mid-season plots one year and that was a mistake. The plants they didn't dig up, they managed to flatten, during ecstatic dust baths that they seemed especially inclined to take at the base of the healthiest, most productive plants. They are now restricted to forays in the early spring and late fall, when there is nothing planted that I want to save.

Two structures that maximize the wonderful ability chickens have to enhance growing conditions are the "henposter" and the chicken tractor. A henposter is simply an area enclosed with a low wall that allows chickens access to all the garden waste and kitchen scraps that are dumped within its perimeter. Our henposter is about four feet square and about eighteen inches high. The chickens think it's the best fast-food hangout in Cluck-Cluck County and spend many happy hours within its confines pecking and scratching and pooping and in fact producing the best compost imaginable, in a very short time.

✳ Dog thinks, "If I ignore them they usually go away."

The soil in the rest of the chicken compound is also astoundingly fertile. A couple of years ago I dug a drainage ditch and piled the soil off to one side of a garden bed. Later, as I was thinning squash plants, I threw a handful of wannabe seedlings over my shoulder and they happened to land on the pile of soil removed from the chicken coop. Somehow that clump of seedlings managed to root and they became the dominant plants in the squash bed, nicknamed Stealth Squash due to their habit of snagging ankles with the excess vine growth that would creep onto the pathways overnight. I believe it is the constant aeration, due to chickens' love for scratching the ground, combined with their ongoing manuring, that creates such wonderful fertility.

The chicken tractor also makes good use of chickens. A chicken tractor is a small light weight plough that is pulled by a team of six to eight chickens.... Oooops, sorry! I just couldn't resist. Seriously, a chicken tractor is simply moveable chicken accommodation, a bit like a gypsy caravan, with its own backyard attached. It has a couple of wheels on one end and handles on the other and no, the chickens don't pull it from one location to the next, their humans do. Chicken tractors can be just about any size, but are usually about four feet wide and eight to ten feet long. Unless you have a "real" tractor or a lot of very strong friends, they mustn't be too heavy. The trick is to build them big enough to house several chickens comfortably, and strong enough to keep predators out, but light enough to move without too much effort. They are excellent parked in a proposed new garden site or over a difficult weed patch that needs to be tamed. In this way the chickens help conserve energy, in this case mine, by doing the work for me.

Our chicken tractor is designed so that it can be placed over dormant and pre-dormant garden beds. It is also the summer home for our leghorn flock, conveniently keeping the two breeds separated during breeding season. This speaks in a small way to another permaculture principle. Elements can and should be designed to be multi-functional within any system. Sometimes this can require something as simple as thoughtful placement.

Well considered placement leads to yet another of the essential elements of permaculture: the sector map. Assuming that a fairly accurate map of the property has already been drawn, the sector map ensures that any new elements being introduced will be installed in the optimum location, where they might in fact serve more than their one intended purpose. For instance, strategic positioning of a shed or clump of bushes can create a useful microclimate. The south wall of the shed can also form the support for espaliered fruit trees or grape vines and further, if the shed is painted white it will help to reflect heat and light back onto the plants close by. (See Albedo Effect in the Glossary.)

Vines planted on the north side of a shed in a windy location will probably not produce any fruit, whereas in the microclimate described above those same vines can be expected to produce a good harvest. This is why it's so important to drawn up a zone and sector map before any serious changes are made. It's mostly just common sense and certainly not that complicated, and yet this is one of the cardinal rules of permaculture design that I instinctively wanted to avoid. Why? I don't know. It's really not that difficult to determine compass points, prevailing wind and water flow, and sun angles for both summer and winter. This invaluable information can be drawn on an overlay placed on top of a property map

✳ Leghorns in the chicken tractor.

showing elements already present on the land, such as dwellings, outbuildings, pathways, etc.

And how did we get from chickens to zones and sector maps? Easy. Just a hop and a skip! These maps will indicate where best to locate, among other things, the chicken run. Close enough to facilitate twice daily visits (zone two or three) but not close enough to draw vermin to the heart/house (zone one); preferably not upwind but within crowing distance, because surely the call of a rooster welcoming the dawn sounds sweeter than the harsh electronic beep of a preset alarm? This just goes to show how tightly integrated and non-linear things are in Permieville, because of course we can just as easily circle back and let the ladies of the Cluck-Cluck County laying-circle have the final words of praise for their prodigious output.

And what does one do with all the eggs? In the spring and early summer, with the ducks and the turkeys as well as the chickens all laying to capacity, the surplus can be overwhelming. We sell at the local market, and we pickle them and curry them and make amazing spinach salads, and potato salads and quiches and frittatas and soufflés and custards and sauces and biscotti and the list goes on...

✳ An egg is such a perfect and versatile food that even in excess they never go to waste.

✳ A "Sunflower Pie," or portobello-tomato frittata, is equally delicious by any name.

There's something very beautiful about a fresh laid egg, still warm from the nest box. It stirs my soul to simply hold one in my hand and after several years of tending hens I still find myself thinking *"This is amazing. What a gift!"* Living in tune with nature, as permaculture has us do, really does help to put the true magnificence of this world into clearer focus.

Ethics and Principles

Before I started to write this chapter I envisioned a neat well-ordered series of paragraphs describing the principles of permaculture. This is not how they are organized in my understanding but I thought a quick scan of my reference material and course notes would help me to distill and regiment the principles into a series of concise paragraphs, each a gift box of information. Not so!

As I blew the dust off several of the permaculture books in my library I began to realize that this was not going to be an easy chapter to write. Just as I remembered, the "wall" of principles still might seem daunting, if not impenetrable, to the uninitiated. And yes, yet again I was thankful that I was in some ways an accidental permaculturist, already walking the walk in many ways, before having any need to talk the talk. What I wasn't expecting was the realization that the principles seem to have morphed and grown in number since the original nine were set down by Bill Mollison in his world-changing manifesto, forty years ago. The additional principles, which swell the number to eleven or twelve, appear to be equally valid. Descriptions of each principle can vary ever so slightly, but at times enough to shift the focus a couple of degrees or more. The ethics of permaculture, which constitute the moral code underlying permaculture, have also developed a somewhat amoebic nature.

This all makes perfect sense considering that the permaculture system is based on closed but naturally evolving cycles that must support and integrate with each other, and to do this must also be able to accept and respond to feedback, so of course the system can be expected to change and expand as it matures.

Ethics of Permaculture

These principles are simple enough and have become part of everyday thinking, especially in the past couple of decades:
- Care for the Earth
- Care for its People
- Equitable Sharing
- Reduced Consumption

Principles of Permaculture

These elaborate on how the aspirations expressed in the ethics can be accomplished.

I didn't envision starting my list with the following two principles but having already referred to the essence of what drives them (the need for systems to be able to integrate with each other and to respond to feedback) in the previous paragraph, I'll let them lead on. To some extent they demonstrate just how true to itself the permaculture system is—shifting, growing, transitioning, all due to feedback.

1. Feedback loops: accepting and responding to change

If I plant the same crop for several years and it never thrives it's telling me that it's not happy. Perhaps I need to adjust the soil, plant a windbreak or accept the fact that bananas don't thrive in northern climes, no matter what.

If I can never supply the demand for eggs at the local market I need to accommodate more chickens.

If my springtime swamp turns into midsummer desert I need

to construct swales and berms to capture and store the excess moisture of spring run-off, which by summer will become a precious and much needed resource.

For a truly integrated system, every "closed" cycle must support and be supported by other cycles. Nothing in the natural world exists as a totally separate entity. This helps to explain my difficulty in deciding which principles should top the list. They are all so interdependent that with even one absent the whole system could eventually fail. Therefore it's difficult to give any one principle precedence over the others.

2. Integrated symbiotic support between all systems: every system must support other systems and in turn be supported by other systems.

Chickens love to eat kitchen scraps and garden waste. They produce the manure that helps grow the garden waste and kitchen scraps.

✳ Goats will consume unwanted shrubbery in double quick time.

Goats are wonderful at clearing out young spruce, which will provide space for a food forest which will in turn provide an excess of the raspberry canes they love so much.

Mycelia, the fine white filaments (hyphae) of fungi, contribute greatly to the breakdown of wood detritus. They are highly instrumental in the natural decay process and plant succession which is part of the forest cycle, and are also invaluable in the breakdown and release of nutrients in organic matter. In return, they use their host as a nursery to proliferate in.

> Fungal mycelia are an amazing entity in their own right, silently performing a wide variety of tasks such as toxic cleanup and pest prevention. They are not understood or lauded half as much as they deserve.

3. Cultivate local species:
avoid introducing invasive species

When considering the integration of all systems it's easy to understand this as the next principle, the need to cultivate native species wherever possible and to absolutely avoid introducing any invasive species. Not only do introduced species disrupt the natural balance of things, those imported from more temperate climates will doubtless also require more tending, which in turn will result in excessive use of resources. Seed brought from a more temperate zone cannot be expected to flourish when introduced to harsher conditions.

Here is also a good place to suggest the cultivation of heritage rather than hybrid plants. Heritage plants were tried and trusted by our forefathers. It was imperative that they produced enough of a yield to survive on (there was no trotting off to the food mart when supplies ran out) plus enough seed for next year's plantings. They learned through necessity to produce hardy, high yielding, low maintenance plants and that's why using local heritage seed,

✳ This beautiful floral motif seeded itself and flourished all summer, after I had tried (and failed miserably) to grow Freesia and Anemones in the same spot. A lesson learned!

from varieties developed specifically for local conditions, makes so much sense.

Generally speaking heritage seeds will produce viable seed which can be saved and used to generate subsequent crops. Hybrid plants, that is, plants that have been developed to enhance certain characteristics, such as reduced number of days to harvest—often to the detriment of other equally important characteristics, such as drought tolerance and disease resistance—do not regenerate true to form. Seed from hybrid plants is often sterile and generally unreliable.

The envisioned order of my list is now totally upended because the mention of yield leads into the next principle.

4. Ensure the fair distribution of yield and empower others to become self-sustaining

No man is an island, trite but true. We are not above or apart from the rest of nature, but just another integrated system that happens to have opposable thumbs and an advanced (sometimes) aptitude for analytical thought. The system we belong to is called community and this community can be identified as both local, and

✳ Sharing our excess at the local community market.

global. As long as one section of our community suffers we are all part of a sickening body.

Statistics show that forty percent of the consumable food in North America goes to waste. That is a heartbreaking statistic, especially considering how many people go to bed hungry every night, and it underlines how essential this principle is.

It feels like time for a little anecdotal ramble, a sort of text-based equivalent of a walkabout. The next principle seems so passive that it risks getting discounted, whereas in fact it might be the most important one of all, so I'm going to push it forward and elaborate.

5. Continuous and mindful observation

I use the term "mindful" because I believe it is necessary to connect on every level, using all our senses in order to really know the land we intend to become part of.

I remember returning from a tax sale after successfully bidding on the abandoned lot adjoining ours. It was a dismal, foggy day and an insidious light rain quickly numbed our fingers as we caressed damp moss on dead branches. We were ecstatic as we clambered and crawled around the derelict lot; through, over and around the tangle of old spruce growth. Although we didn't know it at the time, the lot was destined to become a goat paddock and a food forest, and this year also provided a generous crop of squash and pumpkins.

In those first heady moments of new ownership we were too entranced, too busy connecting at some deep soul level, to dream such dreams, but unbeknownst to us we were taking the first step towards them. We were bonding. Was "mindful observation" enough to get us where we wanted to go? Absolutely not. Accurate, clinical analysis is equally important but the initial bonding is indispensable.

On any terrain the existing natural flora is a very useful indicator of soil type and local conditions, and it doesn't require a degree in botany to access this information. Simple examples: cranberries indicate boggy with full sunshine, whereas ferns like it wet but shady; wild strawberries like sandy soil with full sun while wood asters prefer humus rich soil and shade, etc.

Walking through and glancing here and there is definitely not observing. Only the tip of a massive iceberg of information is revealed in this way. Sitting in various locations regularly for extended periods, that's observing. Short term, this might reveal poor drainage if your butt gets soaked almost immediately or particularly dry conditions if you happen to sit on an ants' nest (not advisable). After such initial contacts, perhaps five plants will become noticeable and then seven and then eleven and insects will appear, all engrossed in their daily tasks, which generally speaking won't involve you.

✳ This attractive wild plant, Kalmia angustifolia, lives up to its more common name, Lambkill, by being deadly to sheep and goats.

A field guide on the flora in your area is indispensable to the process of getting to know the land. Some plants are highly specific in their requirements. For instance I have only ever found Mayflowers (the shy, sweet scented representatives of the province of Nova Scotia) on moss covered banks that face east. This would suggest that they like moist but well drained soil and limited sunshine.

The importance of knowing exactly what is growing naturally on the land was underlined for us when we discovered a patch of Lambkill where we intended to establish a new goat run. This compact shrub has waxy leaves and produces a very pretty pink blossom which totally belies the fact that it is deadly for sheep and goats. Its presence also indicates wet, acidic soil.

The principle of direct observation eases the way to the next principle.

6. Intelligent design and the
observation of naturally occurring patterns

As well as soil conditions, ongoing observation will reveal: naturally occurring microclimates (the temperature in our back lot is consistently about 10°F (5°C) higher than the front); orientation as it affects the availability of sunlight; the effects of prevailing winds; and traffic flow. Deer or rabbit droppings might not seem highly significant but they probably indicate a trackway, and if that trackway transects a veggie plot they also herald a green bean apocalypse. Wild creatures have set patterns and they don't respect detour signs. Some battles aren't worth fighting and it's wise to avoid certain defeat by honouring existing trackways and locating the veggie plots elsewhere.

Patterns are not limited to creatures; they are interwoven into everything. On the deep end are fractals and Mandelbrot patterns, or more familiarly the "golden spiral" of the nautilus shell, dictated by Fibonacci Numbers and the Golden Ratio, the proportion of perfect balance as used by Leonardo da Vinci in such masterpieces as The Last Supper. Water still flows in exactly the same way as it did in the days of this great master and his observations of how water flows can underscore the planning of zones by any winding river or stream. Because the convex and concave banks of any river bend influence the speed of the water that flows by them, the convex side is likely to be scraped clear while the other side is receiving fertile deposits washed downstream and dropped as the current slows down on the concave side of the curve.

As described above, patterns can be as complex as you want to make them, depending on how deeply they are studied. For the purposes of this book and yes, for my sanity, I'm keeping it really simple—except for the previous paragraph!

During WW II code-breakers would search for patterns occurring in any given communication. These patterns became keys to unlock the code and reveal what was really being said. Similarly, naturally occurring patterns can reveal how finely tuned, self-sustaining systems actually work. In their simplest forms, patterns can be called a vocabulary of design.

Plants, for example, have their own specific designs or patterns, as expressed in the placement and shape of their leaves and root systems. This is how field guides typically identify specific plants. The positioning of leaves will be described as alternate, opposite or whorled; and the shape of individual leaves as entire, serrated, lobed or compound.

This is just one example of how everything in nature has its own pattern. The more carefully they are observed the more complex they become, providing an astounding, and highly efficient way of understanding how beautifully choreographed the natural world is.

7. Capturing and storing energy and the efficient use of resources

Capturing and storing energy seems over time to have developed into a principle in its own right but in the original Mollison manifesto I believe it all came under the umbrella heading of Efficient

Use. Certainly there are differences but for simplicity I'm going to lump them all together. Energy can take many forms and as momentum increases down the slippery slope of post-peak oil some effort is being made to conserve and limit the use of fossil fuels, while at the same time ecologically devastating extraction, refinement and transportation processes keep ratcheting upwards. It's heartbreaking to think that the greed of multinationals may well destroy this beautiful planet, but then I'm sure that readers of this book are already aware of all this, and also of the fact that fossil fuels are only one source of energy. For the most part other sources such as wind, sun and water are given little more than a condescending nod once in a while, and are seen at very best as a contributing backup source.

As mentioned in Chapter One, I started capturing and storing water to fend off leaks in the derelict cabin I once wintered in and this led to the installation of a cistern in my new dwelling. The ocean vista was what prompted the southern orientation of the house, which happily enables an efficient passive solar effect. On a sunny winter's day it often becomes too hot to run the wood stove and to increase this effect I could dig a pond out front which would bounce even more of the sun's warming rays through the windows.

Water also has the ability to gather and retain heat. In the greenhouse black plastic barrels of water capture heat through the day that is slowly released through the cooler, evening hours. A stone wall behind a wood stove or by a window will behave in a similar way, by absorbing heat from the initial source (stove or sunlight) and releasing it slowly as the house cools throughout the night.

Most of this is fairly common knowledge in eco-savvy circles. Equally important but perhaps less considered is the energy we invest in any given project. As I write this on a blustery winter's day I'm nibbling on some homemade bread, toasted and slathered with homemade blackcurrant jam. As if on a magic carpet I'm whisked away back to the day I harvested the currants. On

✳ Various forms of energy are captured in these jars of blackcurrant jam.

my knees with the sun filtering through branches weighed down with an overabundant harvest, I was overcome by the beauty and generosity of nature.

This jam has not only captured and stored the energy and benevolence of sunlight invested in the growth and ripening of the currants, the energy of the chickens who provided mulch and fertilizer with their used bedding, my energy spent tending the bushes and making the jam (and sad to say the fossil fuel used to cook the jam) and the essential blessing of regular rain. It has also captured and stored moments of sheer bliss, not to be discounted, especially in harsher times.

It's important not to forget embedded energy. How much of the world's dwindling resources, over and above those physically embodied in the item itself, have gone into manufacturing a simple item such as a flashlight or a coffee maker? The carbon footprint of manufacturing and shipping an item is also an important consideration when calculating embedded energy. This underlines why it makes sense to buy used, recycled items, instead of new. And don't get me started on packaging. I refuse to buy items with excessive and unnecessary packaging, especially as I notice a growing trend among the marketing gurus to make the simplest item appear somehow grander, more elaborate than it actually is, by emphasizing the packaging.

Another energy source worth considering is embedded in the acquisition of knowledge. Time dedicated to learning and education is wasted if it only enhances one existence, when that same knowledge shared could benefit many. This links directly back to the principle of fair distribution and the empowerment of others. It would be fun to assess our homes and in fact our lives listing just how many kinds of energy can practically be captured, stored and shared.

Energy is also recyclable. A grey water system uses water twice, once in the shower and then in the garden. Clothing handed down or modified into quilts and rugs recycles the energy from all the various sources it took to initially produce the item, right back to the shepherd who tended the sheep that produced the wool.

I remember my mother would never waste a piece of bread because of all the energy that went into producing it and that mindset needs to be expanded to include all our resources. Ideally there should be no waste.

I'm going to skate through the next few principles because to me a shorter list is most always preferable to a long one. I feel comfortable doing this because I feel the essence of the following

✳ Garlic pesto is a secondary or "value-added" yield of the garlic harvest; healthful and delicious, it sells well at the market and creates a source of revenue.

principles is either so much part of today's eco-savvy mindset as to be obvious or has already been incorporated into discussion on a previously mentioned principle.

8. Ensure a yield

This principle aligns with efficient use of resources. To squander precious resources for no purpose makes no sense. When I use the resource of my energy I like to have something to show for my time. Similarly, the acquisition of knowledge has no purpose if that knowledge is not put to good use.

9. Start small and move slowly

In a way this principle also aligns with the efficient management of resources. The bigger things get the more energy is required to develop and maintain them and in nature things don't change overnight, not without the help of a natural or man-made disaster. Aggressive plowing and chemical fertilization of fallow ground creates its own disaster zone, disguised as it might be under a field of genetically modified monocrop. Pastured livestock followed up by a few pigs and a healthy dressing of organic mulch will take a year or so longer, but will result in a healthy field with ecosystems thriving in soil that won't blow away.

As megafarms race towards their own collapse it is essential that consumers support local farms, where the various elements of mixed farming support the whole—and monocropping is a dirty word. In so doing consumers also support their local stewards of the earth, who move slowly and respectfully on the limited acreage they have been blessed to farm.

10. Introduce renewable, biological resources only

This is one of those blindingly obvious principles that does not need any elaboration. I only include it because it was on the original Mollison list of principles, and forty years ago, when farmers

✳ Definitely renewable. There's always a barn that needs shovelling out! Hard work, yes, but it sure beats planting with a couple of bags of chemicals.

and gardeners were happily spreading chemical fertilizers on their land, it needed to be said. But wait, what am I saying, isn't that what agribusinesses are still doing on those vast monocropped oceans of genetically modified grain?

This can be a difficult principle to strictly adhere to but if an item such as a plastic barrel for rainwater storage is needed, at least it should be a recycled barrel. Care should be taken to ensure that recycled containers were only used to transport food safe items, such as fruit concentrate, before using them for water storage.

11. Celebrate and value diversity

This applies to cultural and biological diversity. We can learn so much from that which is different from us. Diversity is what enriches every aspect of our lives. For instance, I hate to think how bland, not to mention unhealthy, my typically British diet would be without the enrichment of a world food menu. Xenophobia is indeed a crippling disease. Monocropping is similarly lethal. Any patch of wild zone will testify to the fact that diversity is celebrated and is in fact essential for any system to thrive.

12. See creative solutions not problems

I notice in more recent writings this has been dropped from some lists of principles. It was on the original list and, as it is certainly one of my favourites, it's staying on mine. To borrow from a famous Mollison statement: I didn't have a slug problem, I had a scarcity of ducks. Of course, he was absolutely right. This way of thinking can be applied to all areas of life and it really does make life even more exciting.

Harvest Time:
The Reason for it All!

Creating a sustainable environment such as I have been describing has its own anthem, which might sound very much like the Seven Dwarfs' song in Snow White because, yes, there is always work to be done. How much, or better still, how little depends largely on how well we have thought through and set up each process. For instance, mulched gardens require almost no weeding. Well, that's a no-brainer, but there are more subtle influences that affect the amount of effort required. Perhaps the most significant of these is the season and that is, of course, not very subtle at all. To my mind the fall is probably the busiest time. This might seem like a strange statement given that gardening is usually seen as a summer activity.

Fall is harvest and preserving time. Fall is seed saving time. Fall winterizing ensures that hardy greens, leeks and roots will still be available to harvest well past the shortest day. Fall is especially the time to nourish and nurture the soil, thereby ensuring that during the winter months infinite numbers of microbes and bacteria have an abundance of organic material to work their magic on. The more that is done in the fall, the less that is required in the spring.

✳ A typical sight on the kitchen counter in the fall.

All of the above activities carry equal weight in the cycle of food production. If I don't save seed I have nothing to plant in the spring; if I don't preserve the harvest and therefore have no food, I have no strength to plant in the spring; if I don't nurture my soil it will have no reserves to produce a fine harvest, and so on. Of course, I can go to the store and bring in foreign seed or food, but that breaks the true cycle of sustainability and moves away from one of permaculture's main goals, which is to produce systems which are self-sustaining and do not require input from elsewhere.

The natural forest is a perfect example of this and, considering the size trees can grow to in a natural state, it clearly demonstrates how perfectly our beautiful world is designed. Of course trees die and they fall, but then they rot and provide sustenance for more trees. Even the tiniest of lichens has an important part to play in the sustenance of the forest and the influence of the countless mycelia hidden away underground is truly astounding. (It's definitely worth reading *Mycelium Running* by Paul Stamets.) This is just one example of the intricacies of system interrelationships. Permaculture lore emphasises that nothing exists in isolation. The balance of these interrelationships is highly complex and also extremely fragile. That's why it's so important to honor and learn from all that is naturally occurring around us. Observe, observe, observe—another important Permie premise.

Back to the seasons and especially the busiest one, fall. Where to start? The activities seem to be more varied than at other times of the year. Seed saving, for instance, is quite dissimilar to shovelling out the poultry sheds and if for no other reason, that's a good enough place to start. Seed saving is simple and cost effective,

✱ This commercially made sieve separates the seed from the chaff. A colander and fine kitchen sieve work almost as well.

in that it costs nothing and saves a bundle. Once you have saved some seed and watched it proliferate year after year, it will seem like a complete mystery as to why seed companies continue to flourish.

Some seeds are easier to save than others in that they are numerous, clearly identifiable and easily separated from whatever remains of their flower. Dill is a good example of seed that is uber easy to collect. Those same seeds that flavor dressings and pickles will grow into more dill plants if introduced to soil and moisture. Check out the number of seeds in a bunch of fresh dill. If only a couple of those flower heads were saved (and dried) and if only half the seeds germinated, there would still be more than enough dill growing in the spring. Furthermore, consider the number of seeds in a typical packet of dill seeds and the price per packet. Suddenly seed saving begins to make so much sense.

Pillowcases are not just for pillows. They make wonderful containers for larger bundles of seed heads while they dry. It's quite simple, place a bundle of seed heads inside a pillowcase and hang it out of the way in a cool dry place. As the flower heads dry the seeds drop out and are collected in the bottom of the pillowcase. Various sieves are available for separating any chaff and stem from the actual seeds. Pill bottles are good for saving seeds, as are envelopes. I have yet to formulate a definitive opinion of whether sealed plastic baggies are good or not. They can eliminate any threat of dampness, but only if the seed is thoroughly dry to begin with. And my gut hunch is that seeds like to breathe a bit; after all, they are "living" entities, destined to flourish into exuberant displays of fecundity.

What is essential is labelling. Seed heads seem so clearly identifiable when harvested that several times I have neglected to label large bundles, feeling certain I'd remember what they are. After a month or so of drying, when they're withered and crumbling, they can be much less easily identifiable. Parsnip seeds can look very much like dill for instance. I might have already mentioned that I seem to learn best the hard way.

✳ The small black seeds from the Nigella plants (foreground) are easy to gather, both for kitchen use as black cumin and also for seeding next year's plants.

Now, I label everything, immediately! Also, it is important to put the year on any saved seed because chances are there will be more than enough to cover a year's planting needs. Some seeds, such as spinach and chard, will remain viable for years, just as long as they are kept good and dry; but others, such as parsnip, deteriorate rapidly. I haven't done an in-depth study on the shelf life of various seeds, so I try not to keep seeds for more than a couple of years. However, crop failure, and yes that happens, can result in no seed at all so I don't usually discard any seeds until I have replacements. One additional benefit of seed saving is seed trading and a local "Seedy Saturday" event is well worth a visit. Most seeds are swapped with no money exchanging hands and it's refreshing to spend a little time meeting others of like mind.

Beans and peas are very easy to dry and save, which makes the silly mistake I made this year even more annoying. I planted several kinds of each in the spring and neglected to keep track of

✳ It doesn't take much to share seeds. I like to use the labels as a way to promote the QuackaDoodle blog site. Some people simply present a bowl of seeds, a scoop and an envelope, which is equally fine. Seed swapping is gift giving; it must be fun, never a chore.

which is which. Yes, I know a pea from a bean but early from late variety, or pole from bush? That's where it gets tricky. Some are easy enough to identify such as Mennonite Purple Stripe beans, but other varieties such as snow peas and snap peas are more difficult to identify. They both tasted great and somewhat different while we were eating them, but now they all look just like shrivelled up peas and beans. Hmmmm! Pick each variety separately and label immediately. You'll like yourself much better in the spring, Jenni. So says my inner organizer.

Although squash produce loads of seeds which are easy to separate and save, they have an annoying habit of cross-pollinating and the strain becomes less pure or, totally mongrelled. This can be avoided in a very large garden by keeping the various strains of squash separated by as much distance as possible. Another way is to encase female blossoms in bags and hand pollinate using pollen from a male flower and a sable brush, then only saving seed from the fruit of these flowers. I have never done this because I know such organizational requirements far exceed my capabilities. As a result, I will have to purchase fresh seed next year because all my squash are having identity crises this year.

Some plants are biennials, which means they don't produce seed until their second year. This means they must winter over, which can be tricky in places that experience a hard freeze. Heavy mulching can help, but with crops such as carrots and beets it can be hit and miss. I always need to buy carrot and beet seed. Parsnips on the other hand always put on a great show in their second year, as do Brussels sprouts. Other plants, such as tomatoes and cantaloupe, which I believe are both classified as fruits, protect their seeds with a surrounding of pulp. This needs to be soaked away in a jar of water until the pulp ferments and the seeds can be easily separated from the residue and then dried. This year I was delighted to discover some old tomato seed we had preserved this way and ecstatic to discover that it was still good and able to provide more than enough of a delicious favourite heritage variety that we had lost track of and been unable to obtain elsewhere.

This really helped to underline how important the practice of seed saving is, not only for purely fiscal reasons.

Obviously there is lots more to know about seed saving, enough to fill a book or two no doubt, but I hope that this will at least open the door of enquiry. Preserving the harvest is a much bigger topic and I hope to scratch the surface with what I have to

✳ Parsnips, when overwintered, attain treelike proportions and produce enough seed to plant several acres. N.B. Parsnip seed is unreliable after year one. Strangely enough this giant root remained quite edible, which was an anomaly, I suspect.

say—even though that hope feels faint in my heart because this is the area in which I feel the least competent.

Perhaps I will try to delegate some of the blame for this inadequacy on the pressure cooker my mother used almost daily when I was a child. It terrified me because of the violence of escaping steam, accompanied by my mother's continual warnings of what would doubtless happen if the lid were ever to be removed before the pressure had abated. Whole families had been obliterated, as I understood it, by such catastrophes. I have managed to conquer many of my childhood phobias but in my mind a pressure canner, if I allowed one in the house, would crouch like a malevolent entity, waiting to emit one final heart-stopping hiss of steam. I'm sure they're wonderful but they're definitely not for me. And to further compound the consequences of my limitations, Steinbeck, that wonderful American writer, clearly suggested that painful death can be caused by something as simple as a jar of canned green beans. It's a long time since I read his masterpiece *East of Eden* but what I remember most clearly is that it was my first encounter with the dreaded word "botulism." That word strikes almost as much fear in my heart as the words "pressure canner" and, as fate has it, it seems that without the one you will surely encounter the other. So goes one of the most far reaching, not to mention implausible, excuses for inadequate harvest preservation that you are likely to encounter. However it is mine, and I'm sticking to it. :-)

Botulism, it seems, is not merely a literary device but a very real threat when canning certain vegetables, and pressure canners are universally recommended. The addition of vinegar or sugar to whatever is being processed minimizes this threat so I do make pickles but not in huge amounts because we are not great pickle eaters. We have our favourites which consist mostly of old standbys like pickled beets, pumpkin pickle and rhubarb/ginger chutney. I also make some blackcurrant jam but once again not much because we prefer honey and maple syrup for sweetness.

Sad to say I also tend not to freeze much in the way of vegetables, primarily because we are not super keen on frozen vegetables.

What I do freeze, mostly squash, chard and spinach, goes into winter soups. I tend to focus most of my preservation efforts on cold storage, which is an area that we are still working out. Root cellars or cold rooms are the key for me. Cold rooms are not as difficult to construct and the trick is to insulate them enough to keep heat out but not so well as to have everything freeze solid. Root cellars used to be built right into most rural homes a couple of generations ago, and now that the wisdom of this is again becoming apparent, they

✳ These leeks, dug in the fall, were still quite happy in February/March when stored in a cold basement.

are coming back into favor. However, they are difficult (but not impossible) to install as a retrofit. I don't have one and feel more inclined to rely on a cold room, even though it is my understanding that the risk of freezing is less in root cellars because they are dug down below the frost line.

Another way of producing food during the colder months is to think seriously about season extension methods such as growing in cold frames and greenhouses. This is where I contradict everything I said about sowing greens directly outside in the spring. A fall planting in the greenhouse allows for fresh greens even in the colder months. I don't have a heated greenhouse so I'm somewhat limited to frost tolerant crops. For example I know I can't expect to grow tomatoes in February no matter what I do to keep the greenhouse as frost-free as possible. There are several things that help to keep the temperature up. Heat sinks are a good one; simply pails, preferably black, of water that will heat up through the day and gradually release heat through the night. Blankets of black plastic also help to contain heat gathered during the daylight hours, and double sheets of row cover will help keep heat from the ground from rising up, up and away.

I'm presently experimenting with a traditional French method that was common in the winter market gardens that fed the city of Paris during the times of horse and carriage. In English this particular method is often called a hotbed. Very hot (totally fresh) manure is used as a base, which is then well covered with benign organic material (leaves and straw) topped with lots of rich fertile soil. I think of the manure as equivalent to in-floor heating, the leaves as the floor and the soil as the living area. I played around with this method last year in one of my cold-frames and it did remarkably well, so this year I've put more effort into setting up some manure heated beds in the greenhouse. I will still need to use winter blankets and heat sinks but now I have the addition of a heat source, the manure, which will hopefully provide that couple of extra degrees warmth, often all that is needed to keep plants alive.

✳ Something as simple as a chicken sandwich satisfies more than a physical need when every component is home grown.

As an experiment in winter storage I'm also going to try moving some beets into the greenhouse, leaving their tops on and laying them individually on a shallow tray of moist earth. Most of the rest of the beet crop is either pickled or dried and cured for cold storage. I'm not putting all my beets in one basket; always a good practice when trying something new.

Experimenting with different methods is really the best way to find out what works best for you. I know that's stating the obvious but it is easy to get caught up in other peoples' successes and be determined to make the same practices work just as well for you. Fact is though, conditions are seldom identical, which is why it is so important to experiment, to construct the perfect design for your particular circumstances. This might well be the hidden reason why permaculture appeals to me so much—it encourages creativity and individuality. No matter how hard some might try to deny it, no matter how forcefully we have been squeezed into one mold or another, we are at heart all wonderfully creative individuals. Permaculture allows us to celebrate this. Enjoy and Prosper!

Glossary

Here is a list of some of the main topics mentioned and some terms which may be unfamiliar. I'm sure it won't be seen as comprehensive because permaculture is such a fascinatingly complex topic that a simple list of all aspects could easily become a book in its own right.

It has been my intent all along to keep things as simple as possible and I'm not going to change that now, even though with every topic I have mentioned I thought of a couple more that I could quite happily ramble on about. I was firm but polite in saying to them, "Not this time little ones, perhaps in the next book." As a result I hope the following list will seem useful but succinct.

Acidic refers to the pH content of the soil, 4.0–6.0 being highly acidic, 6.0–6.9 being slightly acidic, 7.0 being neutral. The pH value regulates the availability of nutrients present in the soil. Nutrients become more or less soluble, therefore more or less available to the plants, depending on the pH value of the soil. Potatoes and rhubarb prefer acidic soil. A natural growth of dandelions indicates acidic soil.

Albedo Effect (or reflection coefficient) is derived from the Latin word meaning whiteness and it refers to the reflecting power of a surface. Not to be underestimated, it is an important consideration in the construction of microclimates and suitable planting areas for sun-loving plants. Simply, a south-facing white wall will reflect sunlight back onto plants growing in front of it and maximize their growth. A black wall, or black plastic mulch, will absorb heat, thereby storing the sun's energy for later re-radiation. Microclimates may make use of both direct reflection (albedo) and/or absorption.

Alkaline is the opposite of acidic. Alkaline soils have a pH value starting at the midpoint of the pH scale, which is 7.0. A large range of vegetables such as beets, carrots, broccoli, leeks, lettuce, onions, prefer a neutral to alkaline soil (7–7.5). A natural growth of clover is an indicator of alkaline soil.

Annual These are plants which complete their growth cycle (sprouting from seed, becoming a mature plant, flowering and producing seed) in one year and will require reseeding the next year. Most common vegetables are annuals.

Berms are most often paired with swales in a system of earthworks designed to conserve water and prevent run-off. Berms are mounds which follow the contour lines of the land and are constructed using dirt from the ditch (or swale) dug on the up-hill side of the berm. Berms are placed to prevent erosion. Often topped with small trees, they can be very fertile due to the extra water supply stored in the adjoining swale.

Brassica is the name for members of the mustard family. They are also described as cruciferous (having four equal petals arranged in a cross). Brussels sprouts, cabbage, collards, broccoli and arugula are a few examples of this large group of plants which generally grow well in northern zones and are noted for their high nutritional value.

Companion Planting takes into account the fact that some plants just don't like being close to each other whereas others will thrive in the presence of another species. The arrangement of friendly groups in the garden bed promotes better growth.

Cycles are part of the natural process. Naturally occurring cycles are usually self-perpetuating and relate to other connecting cycles around them by both supporting and benefiting from them. In permaculture the chicken exemplifies this principle.

Diversity is seen as the key to a successful system. In nature diversity rules, supported in part by the symbiotic relationship between its various elements. Each of the elements gives but also receives, creating a tight web of interactivity and reducing vulnerability.

Dynamic Accumulators are invaluable plants, many of which use their long tap roots to access micro nutrients deep in the ground, returning them to the surface where shallow rooted plants can benefit from them. So called "weeds" are some of the best dynamic accumulators, with the much maligned dandelion topping the list by returning sodium, silicon, magnesium, calcium, potassium, phosphorous, iron and copper. Nettles are a close second, returning seven of these precious elements to the soil, and comfrey comes next with a count of six.

Nitrogen fixers are also dynamic accumulators. Specific bacteria colonize the roots of nitrogen fixers (appearing like clusters of small nodules attached to the roots) and change the existing nitrogen into a form readily accessible to plants.

Edges are considered to be extremely dynamic as they allow access to two or more overlapping cycles or zones, each with their own specific benefits, providing the best of both worlds. Maximizing edges is often achieved by meandering rather than straight-edged garden beds.

Energy in all its forms must be conserved. It's as simple as that. Hence the permaculture principle of less for more or increased yield for reduced expenditure.

Espalier is a form of training plants (usually fruits trees) in two dimensional rather than rotund growth patterns. This space-saving technique requires aggressive pruning and is most successful when combined with wall or trellis support.

Food Forests feed people as well as providing natural habitat and windbreak. They carefully mimic natural forest growth by employing the "stacking" arrangement of high canopy, lower canopy, shrubs, herbs, ground cover and mycelium but each of these layers consists of a food producing tree, shrub or herb. A well designed food forest requires little or no input and is fully self-sustaining, just like a natural forest.

Green Manure is a method of improving soil by planting a crop, usually something high in nitrogen, then ploughing it under, or better still cutting it and leaving it in place (the chop and drop method) as a "green" mulch to enrich the soil and add organic material. It can also act as a cover crop, keeping out weeds and protecting soil when land is fallow.

Guilds might be seen as an intensified form of companion planting. Guilds are groups of plants representing the various strata of natural growth (canopy, understory, shrub, groundcover, etc.), carefully chosen for their affinity with surrounding growth and a natural tendency to create symbiotic relationships by supporting neighbouring plants and in turn accepting similar rewards. Guilds are used especially in food forests to establish permanent, supportive networks. Careful study is required to determine the exact requirements and capabilities of various plants as they relate to each other but the effort is well worthwhile. Healthy guilds will remain in perfect balance and highly productive for years.

Heritage in terms of seed stock and livestock refers to strains that have remained unmodified for many generations and have been able to maintain the natural attributes which were valued by our forefathers. By definition this terminology tends to include open pollinated and non-hybrid strains.

Hotbed: this system is used to extend seasons and relies on an underlying bed of "hot" manure to heat soil from below and some form of overlying insulation such as row cover or cloche.

Hugelkultur is a method of preparing a garden bed which mimics natural forest cycles by relying on the decomposition of coarse woody debris to accelerate intense microbial action, thereby heating up the soil and releasing huge amounts of nutrients.

Lasagna Bed is one which is built up on top of the ground using various layers of organic material such as manure, compost, leaves and straw; then topped with soil. This system is very useful in areas of poor soil as it gradually rots down into very fertile soil with high organic content.

Macronutrients: the three main macronutrients are nitrogen (N), Phosphorus (P), and potassium (K).

Mandala gardens are best described as groups of round or petal shaped garden beds designed to allow easy access and maximize the benefits of overlapping and extending edges.

Micronutrients are trace elements, occurring naturally in the soil, which are required in minute but carefully balanced proportions to ensure healthy crops. The main ones are calcium, copper, iron, magnesium, manganese, sodium, sulfur and silicon.

Microclimates are small pockets of extraordinary atmospheric conditions that do not reflect the surrounding climate. They can often be artificially constructed with the use of windbreaks and reflective surfaces such as ponds and whitewashed walls.

Micro-organisms are, for the purposes of permaculture, microscopic life forms existing in the soil which help break down organic matter, releasing nutrients for plants to feed on.

Monocropping is a dirty word and is contrary to all permacultural ethics and principles. This practice of planting large areas with a single crop encourages the spread of disease and insect devastation. The nutrient balance of the soil is disrupted as a single crop takes only the nutrients specific to its needs while returning nothing to adjust the imbalance. Monocropping is ultimately a non-sustainable farming method—typically practised by factory farms.

Multifunctional Design promotes the conservation of energy by ensuring that all elements in the system perform more than one function. Careful attention to placement with regard to existing elements is required initially but this provides an ongoing wealth of savings and benefits.

Perennial plants are those which tend to die off in the fall but will come back in the spring of subsequent years. Rhubarb is perennial and Egyptian or "walking" onions are also referred to as "perennial onions."

Plastic Mulch: Plastic, and all things inorganic are not normally included in permaculture plans but sheet wrap salvaged from lumberyards redeems itself as recycled product. It can be very useful for capturing and storing the sun's rays, either to accelerate the warming of soil or at its extreme to kill off noxious weeds before planting.

Polyculture, the inclusion and nurturing of many varieties, is at the heart of all permaculture plans as it echoes the diversity of the natural world. It ensures the proliferation of symbiotic relationships between cultures to the benefit of each individual, whether plant or animal.

Renewable resources: For a system to be truly self-sustaining it cannot rely on external input, especially when that input comes from a non-renewable source. For example, apart from all the other negatives connected with chemical fertilizers, it comes from a non-renewable external source. Composted chicken manure, on the other hand, just keeps on coming from within its system. It is superior in every way. Moreover, chickens also lay eggs, eat bugs and aerate compost. The choice really is a no brainer.

Sectors are defined by invisible lines radiating from the heart center of a property. They delineate wedged shaped areas of naturally occurring effects, such as sun angles for winter and summer, wind direction, trackways and water flow.

Seed Saving and Sharing ensures sustainabilty and is essential both on a local and global scale. It also makes sound economic sense and helps build community relationships.

Sheet Mulching uses a thick layer of organic materials to deter weed growth, to conserve water, and over time to add organic material to the soil and create fertile growing conditions.

The **Spiral Herb Garden** is iconic to permaculture as it saves space by reaching up rather than outwards, creates microclimates with wet, dry, sunny and shady areas, and addresses the individual needs of plants by establishing heat loving, drought resistant herbs at the top, shade loving "thirsty" herbs at the northern base, and so on. It also employs the spiral design element, which is fundamental in the natural world.

Stacking saves space, mimics natural growth patterns and creates symbiotic relationships. A simple stacking arrangement might have a leguminous (nitrogen fixing) vine trained upwards with a semi-shade tolerant, insect repellent herb (such as rue, wormwood or feverfew) at its base, trading insect protection for its own shady microclimate with enhanced nitrogen supply.

Swales are closed-end ditches dug along the contour lines of a slope to capture and store run off water and prevent erosion. They are usually paired with berms and can become very fertile over time.

Zones are envisioned as radiating out from the heart or house (Zone One), in order to efficiently prioritize placement of utilities based on regularity of use. Kitchen gardens will be close (Zone One or Two) in order to facilitate daily use whereas food forest and wild zones, which require little or no care, will be furthest away in Zone Five.

Suggested Reading

As stated in the introduction, I wrote this book as a primer, as a way to ease (or to lure) readers into the wonderful world of permaculture. It's a beginners' guide, designed to facilitate, not to intimidate. As Permie plans develop more information will surely be sought after. Since the start of our journey Calum and I have learned much from the many highly informative books written about permaculture and related topics. Although they all deserve their place on our bookshelf there are some that really stand out as being particularly useful. A few of these are listed below.

Gaia's Garden by Toby Hemenway (Chelsea Green)
My all-time favourite. It gives me the information I'm looking for in ways simple enough for me to assimilate.

Introduction to Permaculture by Bill Mollison (Tagari)
Definitely written for climates more temperate than the one we live in, but it's important to have instruction direct from the original source of inspiration.

The Permaculture Handbook: Garden Farming for Town and Country by Peter Bane (New Society)
This book is densely packed with everything you might ever need to know about permaculture.

These first three books give an overview of the whole concept. The following give more in-depth information about various integral components of permaculture.

Holy Shit: Managing Manure to Save Mankind by Gene Logsdon (Chelsea Green)

Who knew it was possible to write a fun read about nothing other than manure! This book is amusing but also full of relevant information.

Let it Rot: The Gardener's Guide to Composting by Stu Campbell (Storey)
Also surprisingly interesting considering the topic, this book clarifies the importance of potato peelings...and everything else that goes into making perfect compost. Very useful, if not essential, for someone like me who really just didn't get it.

Sepp Holzer's Permaculture by Sepp Holzer (Chelsea Green)
I like this book because Holzer dreams big, attempts the near impossible and succeeds. Definitely an encouragement for those of us facing challenges.

Mycelium Running by Paul Stamets (Ten Speed Press)
This book was required reading for my Permaculture Design course. I left it until last, expecting it to be an uphill slog, but wasn't I surprised? It's fascinating and (although not its main intent) eloquently underlines the mysterious interconnectedness of all living things by giving a deeper understanding into the power of minute mycelia to instigate mighty consequences. This is very useful for better understanding the workings of hugelkultur and guilds but is also a magical read for anyone, whether Permie-bound or not.

Raising Rabbits the Modern Way by Bob Bennett (Garden Way/Storey) and *Raising Poultry Successfully* by Will Graves (Williamson)
Both these books have been in print for years but are still considered to be two of the standards. All the information needed to start a breeding program.

Index

About the Author

JENNI BLACKMORE is a farmer, art-
ist and writer who built her house
on a rocky, windswept island off the
coast of Nova Scotia almost twenty-
five years ago and has been stumbling
along the road to self-sufficient living
ever since. As a successful micro-
farmer she produces most of her family's meat, eggs, fruit and vegetables, in
spite of often-challenging conditions.
She is certified as a Permaculture De-

sign Consultant from the Falls Brook Centre in New Brunswick
and has published two fiction titles for children and one for adults.

If you have enjoyed *Permaculture for the Rest of Us*, you might also enjoy other

Books to Build a New Society

Our books provide positive solutions for people who
want to make a difference. We specialize in:

Climate Change ◆ Conscious Community
Conservation & Ecology ◆ Cultural Critique
Education & Parenting ◆ Energy ◆ Food & Gardening
Health & Wellness ◆ Modern Homesteading & Farming
New Economies ◆ Progressive Leadership ◆ Resilience
Social Responsibility ◆ Sustainable Building & Design

New Society Publishers
ENVIRONMENTAL BENEFITS STATEMENT

New Society Publishers has chosen to produce this book on recycled paper made
with 100% post consumer waste, processed chlorine free, and old growth free.

For every 5,000 books printed, New Society saves the following resources:[1]

27	Trees
2,407	Pounds of Solid Waste
2,649	Gallons of Water
3,455	Kilowatt Hours of Electricity
4,376	Pounds of Greenhouse Gases
19	Pounds of HAPs, VOCs, and AOX Combined
7	Cubic Yards of Landfill Space

---

[1]Environmental benefits are calculated based on research done by the Environmental Defense Fund and
other members of the Paper Task Force who study the environmental impacts of the paper industry.

For a full list of NSP's titles, please call 1-800-567-6772 or check out our web site at:

www.newsociety.com

new society
PUBLISHERS